Mary Liu Kao, PhD

Cataloging and Classification for Library Technicians

Second Edition

Pre-publication
REVIEWS,
COMMENTARIES,
EVALUATIONS . . .

"**L**ast fall our LTA program used the first edition of *Cataloging and Classification for Library Technicians* as one of the textbooks for our Introduction to Technical Services course. We were so impressed with this edition that we decided to use it as the only textbook for subsequent classes. The revised edition of the book is very impressive. An additional chapter, 'Cataloging on Computers,' has been added and the explanation on the MARC format is quite good. Students and practicing LTAs will understand the various tags and indicators associated with a MARC record. The chapter contains good examples of books, video recordings, serials, and computer disks cataloged in MARC format. Kao also gives the reader a guide to the commonly used MARC tags, fields, and indicators. There is also information on how to search OCLC, which was not included in the previous edition.

The chapter concerning issues and trends has also been expanded to include outsourcing, cooperation among libraries, and the Dublin Core, which provides a core set of metadata elements that can be used to catalog Internet resources.

Dr. Kao has expanded the trends to include the following: online cataloging will be performed by the smallest libraries, LTAs will be hired to replace librarians to do cataloging, and cataloging departments will be merged into automation departments because of the changing nature of the profession. This may allow LTAs to be cross-trained to perform other tasks in the technical services area."

Karen DeLoatch, MLS, MA
Interim Library Director/
LTA Program Coordinator,
Capital Community College,
Hartford, CT

"**H**aving taught with the previous 1995 edition of Dr. Kao's text, I was pleased to see the new edition come forward. The materials are presented in a clear and sequential fashion that lays the groundwork and provides context for the cataloging process within a library. Library services and the management of collections have changed dramatically with the advent of the digital era, and many libraries are in a period of extended transition providing access to both physical and virtual resources. The new edition of this text addresses these changes directly in a new chapter devoted to computerized cataloging.

The dual strength of this textbook is its attention to the standards and organizational concepts that transcend the technology of the day while bringing forward relevant examples and exercises that demonstrate practical applications for library technicians. The new chapter devoted to cataloging on computers is introduced only after sections outlining descriptive cataloging, subject analysis, and classification have been presented. Example records are well chosen to demonstrate the breadth of physical materials that library technicians will be called on to catalog as well as Internet resources.

The final chapter on issues and trends underscores the dynamic nature of libraries in general, and cataloging in particular, and should lead to lively discussion among students as to the pros and cons of issues such as outsourcing.

I recommend Dr. Kao's new edition as the principal text in a Library Technical Associate program and as a reference tool for library technicians in a working catalog department."

Sharon Quinn Fitzgerald, MSLS
Head of Serials
and Library Web Manager,
Fogler Library,
University of Maine,
Orono

"**T**eachers of cataloging are renowned for the demands they make upon their students. Dr. Mary Kao's book demonstrates that she, a long-time professor of cataloging, is even more demanding of herself. Her book includes everything a student could want. Kao introduces her topic in the context of library organization, the role of technical assistants, and the functions of cataloging and classification. She defines relevant terms near the beginning of every chapter and gives detailed citations to the resources upon which catalogers rely. In addition to the latest editions of old standards like AACR2R, LC, DDC, Sears, and Cutter, she furnishes the URLs of Web sites that complement and update the print classics. Her review questions test whether or not the student has grasped both concept and practice.

Organized to promote learning through its treatment of all issues related to cataloging and classification, and generous in the examples provided, *Cataloging and Classification for Library Technicians* will satisfy the needs of both students and everyday practitioners."

Vincent Juliano, MSLS, MA
Director,
Waterford Public Library,
Connecticut

———∽✦∾———

"**S**imilar to the first edition, Mary Kao's *Cataloging and Classification for Library Technicians*, Second Edition, is a good reference and training tool for cataloging library technicians, with extensive explanation of tools used in cataloging: descriptive cataloging, subject headings (LC and Sears), and

Cataloging and Classification for Library Technicians

Second Edition

HAWORTH Cataloging & Classification
Ruth C. Carter, Senior Editor

New, Recent, and Forthcoming Titles:

Technical Services: A Quarter Century of Change: A Look to the Future by Linda C. Smith and Ruth C. Carter

Cataloging and Classification for Library Technicians, Second Edition by Mary Liu Kao

Introduction to Technical Services for Library Technicians by Mary Liu Kao

Cataloging and Classification for Library Technicians

Second Edition

Mary Liu Kao, PhD

The Haworth Press®
New York • London • Oxford

The Haworth Press, Inc., 10 Alice Street, Binghamton, NY 13904-1580

Cover design by Jennifer M. Gaska.

Library of Congress Cataloging-in-Publication Data

Kao, Mary Liu.
    Cataloging and classification for library technicians / Mary L. Kao.—2nd ed.
        p. cm.
    Includes bibliographical references and index.
    ISBN 0-7890-1062-3 (alk. paper)—ISBN 0-7890-1063-1 (pbk. : alk. paper)
    1. Cataloging—United States. 2. Classification—Books. I.Title

Z693.5.U6 K36 2000
025.3'0973—dc21
                                         00-033548

To my late mother,
Remei Bardina Liu,
who inspired me to be
all that I am today

ABOUT THE AUTHOR

Mary Liu Kao, MLS, MS, PhD, was Director of Library Services and Coordinator of the Library Technology Program at Three Rivers Community Technical College, in Norwich, Connecticut, for more than twenty years. She began teaching cataloging and classification in 1974 and has conducted numerous workshops for librarians and library technicians.

Dr. Kao is now a training consultant at Innovative Interfaces, Inc., in Emeryville, California.

CONTENTS

Preface

With more and more libraries being automated, the nature of the work in the cataloging department has changed tremendously. Cataloging has become more technical and less interpretive, with emphasis on uniformity rather than local variations. Cataloging done on the computer has to be precise, and adherence to all the universally adopted rules needs to occur. In most libraries, the burden of performing this task has been shifted from librarians to library technicians, and it is more important than ever that the library technicians receive good education and training.

For the past twenty years, whenever I taught the course Cataloging and Classification in our Library Technology program, I was confronted with the difficult job of finding a suitable textbook for the course. After searching year after year in vain, I had to face the fact that it just had not been written, at least not that I knew of. A textbook in the field of cataloging and classification for library technicians needed to be written.

Every year, I had to resort to designing the content of the course, planning the order of presentation, and using my own notes to teach the course. This was inconvenient for the students. Many prefer to own a textbook that not only facilitates note taking, but also serves as a base of reference and a permanent resource on the subject. Finally, I decided that a book to serve this purpose was necessary, and with the encouragement of editor Ruth Carter of The Haworth Press, this endeavor came into being.

The book is designed as a textbook for a Cataloging and Classification course for the two-year Library Technology Associate Degree or Certificate Program. To provide students with general background information, the course Introduction to Technical Services is recommended as a prerequisite for the Cataloging and Classification course, which is a three-credit, one-semester course.

This text also will serve as a general reference book for library technicians working in the cataloging department. Students are reminded that this text is not a substitute for all the reference tools needed to perform the job of cataloging. This is an interpretation and explanation of the rules and how they should be applied.

I would like to acknowledge the assistance of my friend Joanne Fontanella. Her suggestions have been most useful. Without her editing skills and encouragement, the task could not have been done so smoothly. I am very grateful to her and thank her for her patience.

Preface to the Second Edition

Much has changed in the field of cataloging and classification since the first edition of this book was published in 1995. New editions of the reference tools have been published with modifications and additions. The library environment has changed as well. Automation has taken over, and even the smallest library is now somewhat computerized. Even if a library is not a member of a consortium, chances are that some kind of stand-alone automation system is in place. Cataloging on the computer in MARC format has become much more common, and, therefore, it is important and necessary for library technicians to learn the ins and outs of cataloging in MARC format, as well as the basic skills of descriptive and subject cataloging.

The second edition follows the format and style of the previous edition. It starts with a general introduction on the topics and continues with an in-depth discussion and explanation of all the reference tools that are needed to perform the task of cataloging. Step-by-step instruction is provided so that the target users of this book, students with no library experience, will gradually learn and understand the essence of each task and will feel confident with it.

Every chapter has been revised and updated with new materials presented to reflect the changes and development of the rules and the new editions of other reference tools. A new chapter on cataloging on computers in MARC format (Chapter 8) has been added. Cataloging Internet materials has been included in this chapter. All examples have been updated or verified in the latest editions of the reference tools discussed. In some places, more examples are inserted to illustrate how the theory is put into practice.

Users of this text must have the original reference tools cited here readily available. For example, although *Anglo-American Cataloguing Rules,* Second Edition, 1998, is discussed in detail in Chapter 4, that publication is required to properly and exactly apply rules of performing descriptive cataloging. The book is intended for beginning students as well as library technicians working in the cataloging department who have little previous training.

Once again, I would like to thank my friend Joanne Fontanella for her patience, editing skills, and encouragement all the way through. I would also like to thank my daughter, Patricia, and son, Christopher, for their assistance, support, and encouragement.

Mary Liu Kao

Chapter 1

Introduction

So, you want to be a library technician? Or, more precisely, you want to be a cataloging assistant, or, perhaps, you want to learn more about cataloging? Before studying the essential details of cataloging and classification, a general background is necessary. To have an overall understanding of the operation of a library, information must be acquired regarding the whole library organization and its separate library functions. It is important to know how cataloging and classification fit into the infrastructure. Before we get to the main topics of cataloging and classification, we need to understand the hierarchy of library personnel. We need to explore questions such as, What is a library technician, or a library technical assistant? What kinds of jobs does a library technician perform? What is the relationship between the library technician and other library staff?

TERMINOLOGY

acquisitions: The process of planning, selecting, ordering, and receiving materials in a library.

automation: Computerization of library functions, such as checking books out by computer, ordering materials from vendors or publishers through connected databases, using the online public access catalog, and using online or CD-ROM databases to retrieve information.

cataloging: The process of organizing library materials and making them accessible to library users. Cataloging work is divided into three parts: descriptive cataloging, subject heading, and classification.

1

circulation: Also called access service, this library function mainly consists of checking in and checking out materials, shelving, shelf reading, and maintenance of shelves. Circulation duties also include the maintenance of databases, such as building a users' database on the computer.

classification: The number or a combination of letters and numbers assigned to a work indicating its subject. The purpose is to have materials of the same subject stand side by side on the shelves for easy browsing.

interlibrary loan: When users request materials not owned by one library, the library borrows them from another library on behalf of the users. The requested materials may be mailed, delivered, faxed, or electronically transmitted to the borrowing library. Many libraries offer free interlibrary loan service; others charge a fee.

library technician: A member of the library staff who is in the middle level of the personnel hierarchy, who supervises clerical and student workers, and who is supervised by librarians.The library technician is also called the library technical assistant, abbreviated as LTA, or para-professional, or library support staff. Though there is no national requirement for this classification, the library technician customarily has an associate degree or certificate in the field of Library Technology.

public services: Sometimes called reader's services, these are duties performed in the library's public area that require some contact between library staff and users. Reference services and programming for children are public services. Circulation used to be considered public service, but because it now involves the maintenance of databases, it may be grouped with Technical Services.

reference: A function in which library staff answer inquiries of the users. Reference personnel instruct users on the use of library materials and facilities and assist them in finding needed materials or information.

reserve: Reserve has two meanings. (1) It refers to materials kept in closed stacks, allowing more users to access such materials for a shorter period of time. (2) It refers to materials kept behind the circulation desk that may be checked out only by the requester.

technical services: Services performed behind the scenes in the library for the convenience of the library users. These activities include, but are not limited to, selecting and ordering materials, cataloging and processing materials, and maintaining the databases.

THE ORGANIZATION OF THE LIBRARY

First, let us explore how libraries are organized. Just as every person is an individual, so every library is organized individually. However, one general pattern emerges as a model. Traditionally, all library functions are placed under two sections: *technical services* and *public services.* Technical services usually include all the behind-the-scenes work, such as acquisitions, cataloging, processing, binding, and book repair. Public services, also called reader's services, usually embrace reference, circulation, reserve, and interlibrary loan. As mentioned earlier, each library is organized a little differently, so interlibrary loan might be categorized by some libraries under technical services. Also, with the increasing use of online catalogs, more libraries are classifying circulation under technical services as well. The line between technical and public services is becoming less defined, and many libraries have abolished the division, grouping together personnel who perform related functions. For example, the duties of library personnel who specialize in fine arts may include selecting and acquiring materials in that subject area, cataloging and processing them, as well as answering related reference questions from the users. Some libraries have merged the cataloging department with acquisitions and serials, along with computer technology, into a department of automation and bibliographic control. Always keep in mind that overlapping functions occur in some areas, depending on the organizational culture and policies of individual libraries. There is no right or wrong way of organizing libraries functions.

The organizational chart shown in Figure 1.1 illustrates the division of functions for most libraries. As seen in the chart, the function of cataloging and classification falls under the technical services di-

FIGURE 1.1. Library Organizational Chart (by Function)

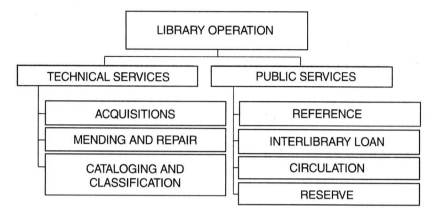

vision of library operation. Some libraries may call this function *bibliographic control*.

THE LIBRARY PERSONNEL

The American Library Association, in its 1976 statement *Library Education and Personnel Utilization*, defines the categories of library personnel and their qualifications. Until now, this was the only official guideline in existence. This document separates library personnel into two categories: professional and supportive. The professional librarian requires a master's degree in Library Science (MLS). The supportive category includes the library associate, the library technician, and the clerks. The minimum requirement for library associate is a bachelor's degree, and for library technician, two years of college-level study, or an associate degree, or a one-year certificate. Clerks are not required to have college education but do need to have clerical skills and in-service training. The chart shown in Figure 1.2 is a convenient way to demonstrate the library personnel hierarchy.

This chart reflects the official guidelines; actual classification may vary from library to library. In many small libraries, for example, the staff classification is purposely vague and there may be only two categories: professional and nonprofessional. Some very small libraries

FIGURE 1.2. Library Organizational Chart (by Personnel)

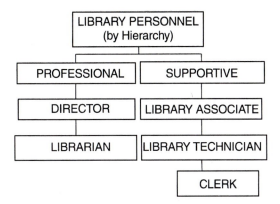

do not even make that distinction, and as a result, everyone who works there is a "librarian." The only nationally recognized requirement is that the librarian have a master's degree in Library Science. Not only is there no universal standard requirement for the next level of library personnel, but there is also no universal agreement on the title for these dedicated, important library staff. For this book, the title *library technician* is chosen. Other publications may use different acceptable titles, such as *library technical assistant (LTA), library support staff, paraprofessional,* or *library assistant.* Used less frequently and generally considered inappropriate, other titles that have appeared in the literature include *nonprofessional, subprofessional, library aide,* and *paralibrarian.* Today, the overall consensus sets the qualification for library technician as a bachelor's degree or an associate degree or certificate, or education and training gained from a Library Technology Program in a four-year or community college. This issue is being discussed extensively in the library world, and there are indications that, in the near future, either an associate degree or a certificate will be necessary to qualify one as a bona fide library technician.

CATALOGING AND THE LIBRARY TECHNICIAN

Before we talk about what types of jobs library technicians perform in the area of cataloging, let us examine what cataloging is and why it is the most important function in the total library operation.

Cataloging is the process of organizing library materials and making them accessible to library users. The challenge the cataloger faces is accommodating the many diverse approaches people use in retrieving library materials. A good cataloger makes it possible for users to find materials easily, whatever approach is applied. Good cataloging practices make library operation more efficient. The most wonderful collection is useless if the materials are not easily accessible to library users. Cataloging is also called bibliographic control, and understanding the why and how of bibliographic control is not only vital to staff working in the cataloging department, it is important to all library personnel. Reference and public service staff are more likely to perform their duties effectively if they have some background knowledge in cataloging and classification. This includes understanding how the collections are arranged for use and how to assist users in finding needed information in an accurate and timely fashion. Special training is necessary to ensure that library staff acquire knowledge and skills on the theories and procedures of the cataloging and classification process. Continuing education involving new developments and trends in the field, including rule changes and revisions, must be available.

To facilitate the cataloging process, guides and codes are designed, reviewed, and redesigned by library organizations. The official guides and codes, in turn, are adopted by libraries to ensure uniformity from library to library. In a time when people move from place to place and use different libraries at different times, consistency is essential. Even for the users of small libraries, confusion is minimized when officially adopted standards are practiced. In the sense of shared catalogs and personal computer connections, automation allows access to many different libraries simultaneously; more reason for uniformity in cataloging and classification rules. Knowledge of cataloging rules and codes, along with a general background in cataloging, encourages all library staff to understand how the collection is organized and to identify quickly and easily the materials contained within the local collection or from other remote-access libraries.

In the area of cataloging, the library technician's job has changed in the past decade chiefly because of automation and budget restraints. Many tasks done in the past by librarians are now routinely performed by library technicians. A modern-day library technician accomplishes almost every task in the cataloging department, usually working at a computer terminal connected to nationwide or area databases. The library technician's job is to retrieve bibliographic information from print sources or computer databases to match the locally acquired materials, and to input local collections into the database. Traditional cataloging support tasks such as typing, duplicating cards, and filing are gradually being eliminated because of automation. Nowadays, library technicians need to possess more sophisticated skills. The library technician is expected to process materials properly and follow through with the procedures that will ensure that materials reach their proper locations on the shelves. Other duties include maintaining a clean database for the online catalog or, if a card catalog is still in use, maintaining the card catalog.

It cannot be overemphasized that proper cataloging is essential for library operation. The Internet, however, provides no legitimate cataloging of data and information. Even though a massive amount of information is available, it is extremely difficult and inefficient to retrieve the exact information that one may need. The tasks that govern the art and science of cataloging involve numerous rules and codes, which may suggest tedious and trivial processes for some. However, this mechanical and precise aspect of cataloging is responsible for the convenient, efficient, and workable system that we now have. The job of cataloging is important, challenging, interesting, and rewarding.

REVIEW QUESTIONS

1. How is the library organized by function?
2. How are library personnel organized?
3. What is the definition of library technician?
4. What are the qualifications for a library technician?
5. Why should libraries adopt the same cataloging rules and codes?
6. Why is cataloging a very important function in the total library operation?

Chapter 2

Library Catalogs

What is a library catalog? A library catalog is a record or a list of the collection of a particular library, or of the collection of many libraries that are connected electronically. When it is a combined list of the holdings of many libraries, it is called a union catalog or a shared catalog. We can also say that a library catalog is an organized list of information resources arranged in logical, prescribed order. Why is it necessary for every library to have a catalog? Catalogs serve many different functions that will be explained in this chapter. Basically, catalogs are established so that library users are able to retrieve the needed information. A good catalog is a good information delivery tool. To produce a good catalog, all materials must be cataloged so that they can be found. Catalogs come in a variety of sizes and formats. When studying about library catalogs, it is necessary to understand what are known as individual entries that identify each item in the collection. As a library technician, you will be required to interpret the entries to the users, if you work in the public services area, or, if you are a cataloger, to actually work on determining how to properly enter information.

A library catalog is never complete because the library collection is a living institution. Materials are added on a daily basis as well as removed at regular intervals. To accurately reflect the collection of the library, it is necessary to update the catalog constantly.

Materials listed in the catalog represent everything the library owns, plus collections from other libraries in the case of a shared catalog. Included in the catalogs are books (also referred to as monographs), periodicals (also referred to as serials, which include both professional journals and popular magazines, newspapers, and other types of publications that are published continuously), pamphlets, audiovisual materials, computer files, and digital information.

TERMINOLOGY

bibliographic record: A term used to describe the cataloging information for an item. Included are author, title, publisher, date, physical description of the item, and any other pertinent information needed to identify the material as a unique item.

book catalog: A listing of the library's holdings in book form. A computer printout is the latest form of a book catalog.

card catalog: A form of catalog that is made up of 3 × 5-inch cards. On each card, information about an item is written, typed, or printed. Cards are arranged in alphabetical order and filed in drawers especially designed for this purpose. The interfiling by author, title, and subject cards is called a dictionary catalog. When cards are filed separately so that all subject cards are filed in one section, and all author and title cards are filed in another, this is called a divided catalog.

CD-ROM catalog: The compact disc read-only memory is an optical disk played on a special disk player linked to a computer terminal. The library's collection is engraved on the disk, and when the disk is played on a CD-ROM drive connected to a computer, information can be retrieved.

COM catalog: Computer Output Microform catalog is a listing of the library's collection that is either on microfilm or in a microfiche format. A microfilm/fiche reader machine is used to read the information.

library catalog: A list or a record of all the materials in a library. May also include materials from other cooperating libraries that belong to the same network or consortium.

OPAC: The Online Public Access Catalog is a listing of the library materials that can be retrieved on a computer terminal.

union catalog: A combined catalog that includes the collection of groups of libraries. There are local union catalogs, nationwide union catalogs, and international union catalogs. A union catalog is also called a shared catalog.

FUNCTIONS OF THE CATALOGS

Why is it so important to have an accurate and up-to-date catalog? Here are the functions of a catalog:

1. To indicate to the users what is housed in the library. Catalogs list every single item acquired by the library. In the case of a union, or shared catalog, besides displaying what the library has, the catalog also shows what the library can obtain for the users.
2. To help users make the proper selection. With all the information in the catalog, users are able to get pertinent facts such as author, title, publisher, publication date, relevant subject, and the format of the material, such as book, video recording, or computer file.
3. To provide access to the materials, whether through the author, title, or subject. Location is indicated by a letter and number symbol referred to as a call number. This letter and number combination indicates exactly where the wanted materials are shelved or stored. In a union or shared catalog, the location column also identifies the name of the library that owns the materials.
4. To function as an indispensable tool for library staff in the areas of acquisition, cataloging, inventory control, and reference works.

TYPES OF CATALOGS

The Book Catalog

The book catalog is the earliest form of catalog. Ancient libraries listed the titles in the collection on papers that were bound in book form. In the nineteenth and twentieth centuries, the card catalog became widely accepted and almost completely replaced the book catalog. In the 1960s and early 1970s, computerized libraries started to print book catalogs again.

Because supplements have to be produced frequently and attached to the existing catalog, the book catalog is inflexible and cumbersome for the users. The advantage of using a book catalog in the form of a computer printout is that new entries are automatically filed, reducing the labor cost for library personnel. Also, many copies can be made available for different locations, such as for branch libraries, for students' dormitories, faculty offices, etc.

The Card Catalog

Since the Library of Congress launched the printing and selling of catalog cards in 1901, the card catalog, up to the late 1980s, has been the most widely used type of catalog. The card catalog uses 3 × 5-inch cards filed in alphabetical order in drawers that fit in a specially designed cabinet. Libraries either type or print their own cards, have an outside printer print the cards, or, more often, buy the already printed cards from the Library of Congress, a commercial book dealer, or one of the many library supply companies.

The card catalog system offers flexibility. New cards are interfiled in their correct order constantly. Cards may be removed easily from the catalog to reflect changing status, such as withdrawal or loss of the item. Cards are relatively inexpensive and easily accessible.

The main disadvantage is filing. It is labor intensive. The library filer has to be very familiar with all the filing rules and work very carefully and competently. A misfiled card represents an item with no reference in the catalog and may be permanently lost. As mentioned earlier, cards need to be filed and removed constantly, and, therefore, maintenance of the card catalog is a burdensome and time-consuming task. With automation, filing is done by the computer, and the mechanical problems of maintaining the card catalog have been solved, rendering the card catalog system obsolete. In the 1980s, many libraries installed computerized catalogs, ceased to file new cards in the card catalog, and often stopped maintaining the card catalog. These frozen catalogs remained temporarily for reference purposes and to hold information on older materials not entered into the database. Finally, when the total collection had been entered into the computer database, the card catalog was given a death sentence and taken away to make room for computer terminals.

COM (Computer Output Microform) Catalog

In this format, bibliographic records are photographed and produced on microfilm or microfiche, which is relatively inexpensive. Space is saved compared to the card catalog and the book catalog formats.

The disadvantages are somewhat similar to those of the book catalog, in the sense that it is difficult and expensive to update. It is inconvenient for users to employ the many supplements and troublesome, initially, to learn to use the necessary equipment, the microfilm/fiche reader/printer. It also means extra expenses because the library has to acquire several of these machines. This form of catalog was adopted by some libraries for a short while in the 1970s but never became very popular.

OPAC (Online Public Access Catalog)

OPACs began to appear in libraries in the late 1970s and the early 1980s. They quickly gained wide acceptance and became the most popular catalog form. With either the touch screen or the keyboard, users can access the most up-to-date information on the library's collection and can get a printout of that information. OPAC offers fast retrieval and an immediate display. In a shared online catalog, users can retrieve information from other participating libraries. These systems not only indicate the holdings of different libraries but also tell the circulation status of an item, whether it is on the shelf and, if not, when the item is due back. Some systems allow users to place a hold on the desired item or to directly request an item from other libraries through interlibrary loan agreements among the libraries in the system.

Due to advances in computer technology and the implementation of standards in the technology world, library users now can search hundreds of online catalogs through remote log-in facilities on the Internet. The new generation of OPACs is easier to use and offers more options.

The online catalog has changed traditional cataloging in several ways:

1. The dependence on shared bibliographic databases for cataloging has increased copy cataloging and decreased original cataloging activities.

2. The trend is toward linking the holdings of one library to other local libraries, to other libraries in the country, or to the international database.

3. The original catalog search methods based on the simple author, title, subject arrangement have become a multitude of approaches, such as subject key word search, title key word search, Boolean search, and search by call number.

4. The library catalog has expanded to include commercially produced reference data, such as index and abstract services, and full-text articles.

5. Electronic, or digital data, including Internet resources, are included in the catalog.

6. The online catalog does not stand alone, and in most libraries, it is an integrated system used for acquisition, circulation, reserve, and record-keeping functions.

7. Users can have remote access to the database through their own personal computers, from homes, offices, dormitories, schools— from anywhere in the world where portable computers have remote access capability.

CD-ROM (Compact Disc Read-Only Memory) Catalog

CD-ROM technology makes it possible to have a library's holdings engraved on computer disks. The cost for an individual library to have its collection put on compact disk used to be restrictive, and as a result, libraries did it collectively with other libraries in the same networking environment. It has become a popular format for library consortia. One CD-ROM disk has a storage capacity that is equivalent to 300,000 printed pages.

The CD-ROM player is now an integral part of computer equipment. Information can be shown on the computer terminal, and for most users, it is indistinguishable from the online catalog.

New disks can be produced easily and quickly to update listings. Because it does not need to connect to external databases, there is no computer downtime problem. Thus, this has become the most popular backup system for the online catalogs. The CD-ROM market has grown rapidly in the 1990s as an inexpensive substitute for the online library system.

The disadvantage of the CD-ROM catalog is that it is not as up-to-date as the online catalog. Because it is not interactive in nature, it does not offer such convenient features as item status, reserve, and interlibrary request, which the online catalog can offer.

ELEMENTS OF A BIBLIOGRAPHIC RECORD

No matter what format the catalog takes, the information displayed is the same. Figures 2.1 and 2.2 show how the same information is displayed both on a catalog card and on the computer screen. Information that a book catalog reveals is also the same since a book catalog is either a computer printout or photographs of catalog cards.

Information in a bibliographic record includes the call number (CB161.A35 1987), the author (the example in Figures 2.1 and 2.2 does not show an author), the title proper, with statement of responsibility (An Agenda for the 21st century/ [compiled by] Rushworth M. Kidder), the edition statement (no edition statement is shown for this book, indicating it is the first edition), the place of publication (Cambridge, Massachusetts), the publisher (MIT Press), the date of publication (1987), the extent of the item, other physical description (xxii, 216 p. : ports. ; 21 cm.), the series title, notes ("The interviews in this book were

FIGURE 2.1. Information on a Catalog Card

CB
161
.A35
1987
	An Agenda for the 21st century /
	   [compiled by] Rushworth M. Kidder.—-
	   Cambridge, Mass. : MIT Press, c1987.
	   xxii, 216 p. : ports. ; 21 cm.
	   Bibliography: p. xxii.
	   Includes index.
	   "The interviews in this book were originally published
	as a series in the Christian Science Monitor"—
	T.p. verso.
	   ISBN 0-262-11128-4
	
	   1. Twenty-first century—Forecasts.
	I. Kidder, Rushworth M. II. Title:
	Agenda for the twenty-first century.

21 DEC 90	16578832	MHGAdc	87-22597

FIGURE 2.2. Catalog Information on a Computer Screen

FORMAT:	book
LOCATION:	BrdgprtPop 303.49 A265f
	CCSU UCStamford CB161 A35 1987
	AsnuntckCC CCSU ThamesVlyC UBridgport
	UCStorrs UCTrecker CB161 .A35 1987
	HartfordPL CB161.A35
	MoheganCC LCC CB161 .A35 1987
	MeridenPL ANF 303.49 AG
	NorwalkPL NorwalkSth SheltonPL
	303.49 AGE
	StratfrdPL 303.49 K46A
	TunxisCC WCSU CB161 .A35 1987
	WrthrsfldPL 303.49 KIDDER
CONTROL NBR:	16578832
LC CARD NBR:	87022597
ISBN:	0262111284
TITLE:	An Agenda for the 21st century / [compiled by] Rushworth M. Kidder.
PUBLISHER:	MIT Press,
DATE:	c1987.
DESCRIPTION:	xxii, 216 p. : ports. ; 21 cm.
NOTES:	Includes index.
NOTES:	"The interviews in this book were originally published as a series in the Christian Science Monitor"—T.p. verso.
NOTES:	Bibliography: p. xxii.
SUBJECT:	Twenty-first century—Forecasts.
CO-AUTHOR:	Kidder, Rushworth M.
OTHER TITLE:	Agenda for the twenty-first century.

originally published as a series in the Christian Science Monitor"—
T.p.verso), the ISBN number (0-262-11128-4), subject headings
(Twenty-first century—Forecasts), and other added entries (Kidder,
Rushworth M.; Agenda for the twenty-first century). The information
on the bottom of the card in Figure 2.1 may be of no concern to the users, but it is a record for the cataloger. Included are the date this book
was cataloged online (21 DEC 90), the computer control number of the

system (16578832), the holding library symbol (MHG Adc), and the Library of Congress Control Number (87-22597). On the online catalog, the format is indicated (book), and the location is listed, not only with the names of the libraries, but also with the individual call numbers as well, to facilitate interlibrary loan procedures. Note that not every bibliographic record lists all the aforementioned features. However, the relevant information is there to assist users in retrieving the desired material.

Figures 2.1 and 2.2 dramatically demonstrate how much information is included in a catalog and how each entry describes a very important feature of the item. It is necessary for the library technician to know how to read the information, how to extrapolate the relevant parts, and how to organize the entries so that gathered information follows a prescribed formula. These procedures, which are explained at length in Chapter 4 of this book, are prescribed in detail in *Anglo-American Cataloguing Rules,* Second Edition, 1998 Revision.

REVIEW QUESTIONS

1. Explain the functions of the catalogs.
2. Explain the different types of library catalogs.
3. What are the advantages and disadvantages of each type of catalog?
4. What information is included on a bibliographic record?

Chapter 3

Tools Used for Cataloging

TERMINOLOGY

call number: This is a combination of a classification number and a letter and number code representing the author. Each call number is unique and enables users to locate the needed material on the shelf—the address of the material.

copy cataloging: The library staff copies or matches the cataloging information that is already completed by another cataloger from another library. The completed information can be found in some print sources, but, most likely, the library technician will get such information from a computer database.

descriptive cataloging: This is the first step of the cataloging process. This means describing the material physically and determining the choice of access points (headings). This is done by following the rules listed in the reference tool *Anglo-American Cataloguing Rules,* Second Edition, 1998 Revision.

original catalog: The library cataloging staff performs all the procedures to completely catalog materials. The tasks include descriptive cataloging and assigning subject headings, classification numbers, and book numbers.

subject cataloging: This second step of the cataloging process is divided into two parts. First is to assign subject headings to the materials by using either *Sears List of Subject Headings* or *Library of Congress Subject Headings,* whichever the particular library chooses. The second part is to assign a classification number to the material by using either *Dewey Decimal Classification Schedules* or *Library of Congress*

Classification Schedules, again depending on the choice of the particular library. A book number is added to the classification number to complete the process. The book number is assigned according to *C. A. Cutter's Three-Figure Author Table*.

subject heading: This term is used in library catalogs to describe the subject matter of the materials and as an added access point for retrieving the materials using the subject approach.

INTRODUCTION

To properly catalog library materials and to maintain consistency, rules need to be followed. It is even more important to follow the rules rigidly when the library joins a network and the catalog becomes part of the shared database. Reference books or the electronic versions of reference books must be acquired and the rules practiced. These references are called cataloging tools or tools for cataloging.

Cataloging is done in two ways: original cataloging and copy cataloging. Original cataloging means that the entire process of cataloging is completed locally by the library staff. In copy cataloging, a shortcut is taken that entails copying the necessary information from a source that contains works already cataloged. Most cataloging can be done by way of copy cataloging, thus avoiding the unnecessary time and effort spent in duplicating the cataloging processes. This is especially true in automated libraries, where the needed cataloging information can be found in the database to which the library has access. On occasion, however, for some very special or unique materials, necessary information for copy cataloging cannot be found. In such cases, original cataloging must be performed. Many odd documents and publications of local interest belong in this category. While copy cataloging is done for the majority of library acquisitions, some materials are always waiting to be cataloged locally. In the past, libraries used the catalog librarian to perform original cataloging, while assigning to the library technician the job of copy cataloging. With the continuous emphasis on better training and more education, many library technicians now perform both jobs.

Steps or stages for performing original cataloging are divided into two parts: descriptive cataloging and subject cataloging. Subject cataloging has two additional stages: subject heading and classification. The chart shown in Figure 3.1 illustrates the steps for cataloging.

Rules and regulations are formulated and published for each step. These rules have been adopted universally by the library world to provide consistency and uniformity for users. The library technician needs to be familiar with appropriate cataloging tools that contain the rules, so that rules for specific material can be followed accurately. Memorizing all the rules is neither possible nor practical, although the library technician needs to understand the rules and how to consult the reference tools efficiently. The following references are used as cataloging tools.

ANGLO-AMERICAN CATALOGUING RULES, *SECOND EDITION, 1998 REVISION*

The *Anglo-American Cataloguing Rules,* Second Edition, 1998 Revision, is used when performing the first step in cataloging, called descriptive cataloging. This revised edition was published jointly by the American Library Association, the Canadian Library Association, and The Library Association (Great Britain) in 1998. Generally referred to as AACR2R, this collection of cataloging rules has been adopted by almost every library in the United States, Great Britain,

FIGURE 3.1. Steps for Cataloging

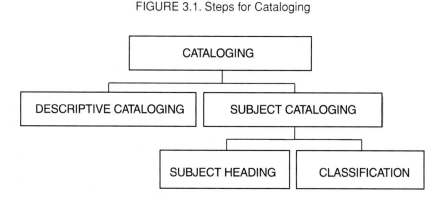

Canada, and Australia. AACR2R supplies rules for the physical description of materials of all formats, including books, pamphlets, printed sheets, cartographic materials, manuscripts, music and sound recordings, motion pictures and video recordings, graphic materials, computer files, three-dimensional artifacts and realia, microform, and serials. It also presents rules for establishing the access points (also referred to as headings or entries) for users to retrieve materials, technically referred to as main and added entries. AACR2R is discussed in detail in Chapter 4 of this text.

An electronic CD-ROM format of AACR2R, called *AACR2R-e* is available for use. AACR2R can also be found on a CD-ROM disk produced by the Library of Congress titled *Cataloger's Desktop*. From this disk, the cataloger can look up rules while cataloging on the same screen. In addition to the contents of AACR2R, *Cataloger's Desktop* also includes many other Library of Congress publications that are used as references when cataloging online: *Library of Congress Rules Interpretations; Subject Cataloging Manual: Classification; Subject Cataloging Manual: Shelflistings; MARC 21 for Bibliographic Data; USMARC Format for Authority Data; USMARC Format for Holdings Data; USMARC Format for Classification Data; USMARC Format for Community Information;* the latest edition of all five *USMARC Code Lists;* plus fifteen other publications in more specific areas. *Cataloger's Desktop* is fully updated quarterly.

LIBRARY OF CONGRESS SUBJECT HEADINGS

The selection and assignment of subject headings to materials is the second step in completing the cataloging process. *Library of Congress Subject Headings,* usually called LCSH for short, is one of the two tools used to perform this task. Updated constantly, LCSH establishes subject terms to be used by catalogers when assigning subject headings for a particular item. Furthermore, it establishes terms related to the subject at hand, plus broader and narrower related terms. Finally, it lists terms that the cataloging staff should not use. Now in its twenty-second edition (1999), LCSH has been adopted and is used by thousands of libraries and a multitude of print indexes.

The print edition of LCSH is updated annually and is available in microfiche format, which is updated quarterly. The electronic version of this publication can be found on a CD-ROM titled *Classification Plus*. *Classification Plus* is a full-text, Windows-based CD-ROM product that contains the *Library of Congress Classification Schedules* and the *Library of Congress Subject Headings*. This version is updated quarterly as well.

LC Subject Headings Weekly Lists is now available only electronically on the Web page <www.lcweb.loc.gov/catdir/cpso/wls.html>. This weekly compilation lists headings that Library of Congress catalogers have created, changed, or deleted.

Details of the *LCSH* and its use are discussed in Chapter 5 of this text.

SEARS LIST OF SUBJECT HEADINGS

Small public libraries and school media centers usually choose to use the *Sears List of Subject Headings* as a handbook to complete the function of assigning subject headings to materials. *Sears* is adopted in most cases to avoid the complexity of the *Library of Congress Subject Headings*. *Sears* lists fewer terms than the LCSH, but those listed are basically the same, with simplification and modification more appropriate to the needs of smaller libraries. As the collection becomes larger and *Sears* proves to be inadequate, libraries may decide to switch to the LCSH. Because of automation and shared cataloging, most networks require their members to use LCSH, and as a result, the use of *Sears* has decreased.

A new edition of *Sears* is published as deemed necessary by the publisher. The latest one, the sixteenth edition, was published in 1997. More on *Sears* and its usage appears in Chapter 5 of this text.

DEWEY DECIMAL CLASSIFICATION AND RELATIVE INDEX

After assigning subject headings, the next step in cataloging is to assign classification numbers so that materials pertaining to the same subject are put side by side on the shelves, making it more convenient for library users to find related materials when browsing the stacks. Though some local classification systems are in use, the two most important classification systems employed in the United States are the Dewey Decimal Classification and the Library of Congress Classification systems. Most small and medium-sized public libraries and virtually all school media centers choose to use the Dewey Decimal Classification system.

The *Dewey Decimal Classification and Relative Index* is published in two editions, full and abridged. The latest full edition is DDC 21, published in 1996 by OCLC/Forest Press. In the publisher's foreword, it is stated that libraries with a small collection of up to 20,000 volumes that do not anticipate significant collection growth may choose the Abridged Edition 13. The DDC is kept up to date between editions through the monthly posting of new and changed entries on the Dewey home page <www.oclc.org>, and through the annual publication of *Dewey Decimal Classification Additions, Notes and Decisions* (DCand).

An electronic version of DDC 21, titled *Dewey for Windows,* is available on CD-ROM from its publisher, the OCLC/Forest Press.

A more thorough discussion of the Dewey Decimal Classification system can be found in Chapter 6 of this text.

LIBRARY OF CONGRESS
CLASSIFICATION SCHEDULES, A TO Z

For libraries not using the Dewey Decimal Classification system—mainly the larger public libraries, special libraries, and academic libraries—the Library of Congress Classification system is the choice. The contents of this system are published as a set of alphabetized, coded paperbacks called class schedules. Currently, the system includes forty-six volumes altogether. Each class schedule is published and revised independently at different times. For example, the latest revision for Schedule A was published in 1998, whereas Schedule G was last published in 1976.

The Library of Congress publishes *LC Classification—Additions and Changes* four times a year to keep catalogers up to date. Both newly added and changed numbers are listed in this publication. Some commercial companies publish class schedules with a few years of *LC Classification—Additions and Changes* incorporated into the main class schedules for the convenience of the catalogers. Gale Research Company also produces a CD-ROM version of the schedules, with their additions and changes, titled *SUPERLCCS on CD-ROM.* Another electronic source for class schedules, a CD-ROM titled *Classification Plus,* includes *Library of Congress Subject Headings* produced by the Library of Congress and is available as an annual subscription with quarterly updates. Not all the schedules can be found on CD-ROM, although the latest edition, Issue 3, 1999, does include twenty-seven classification schedules. Information on all kinds of Library of Congress publications, including the class schedules for both print and electronic versions, can be found on the library's Web site <lcweb.loc.gov>. The *Library of Congress Classification Schedules* and their use is discussed in greater detail in Chapter 6 of this text.

C. A. CUTTER'S THREE-FIGURE AUTHOR TABLE

To create a unique call number for easy identification, a Cutter number, also called a book number, must be added to the Dewey Decimal Classification number. Usually called an author number, the Cutter number facilitates a logical ordering on the shelves. The number is derived from the *C. A. Cutter's Three-Figure Author Table,* or from another edition titled *Cutter-Sanborn Three-Figure Author Table,* or from some other abbreviated versions. Libraries using the Library of Congress Classification number employ a simplified table, the "LC Book Number." Besides the printed version, Cutter numbers can be found on Southern Illinois University Library's Web site <www.lib.siu.edu/swen/cutter.htm>. Author tables and their uses are discussed fully in Chapter 6 of this text.

Now that you are familiar with these cataloging reference tools, you are ready to perform the great act of cataloging! Let us take it one step at a time.

REVIEW QUESTIONS

1. What are the two ways of cataloging?
2. What are the four steps of cataloging?
3. What are the necessary cataloging tools for a library using the Library of Congress Classification system?
4. What are the necessary cataloging tools for a library using the Dewey Decimal Classification system?

Chapter 4

Descriptive Cataloging

access point: A name or a term that can be used to retrieve the bibliographic information from a card catalog or an online catalog. Examples are: author's name, title of the book, and subject heading. All entries, or headings, are access points.

added entry: Other access point(s) besides the main entry used to identify a work. For a book, added entries may include joint author, translator, title, series title, etc.

analytical entry: An access point that is the title or name of a part of a work, or a separate part that belongs to a series.

area: The part of the description that contains certain pieces of information. For example, title and statement of responsibility make up an area; edition statement is an area.

chief source of information: The source the cataloger uses that provides the information for cataloging the material. For example, for a book, the chief source is the title page of the book. For a video recording, the chief source is the title frames or the information printed on its container.

compiler: A person who prepares for the publication of a work by putting pieces of related works together. The pieces may be written by the same author or by several authors.

diorama: A three-dimensional miniature scene, such as figures in a background setting.

editor: A person who did not write or produce the material but is responsible for it. The editor sometimes writes commentaries or an introduction to the work.

explanatory reference: A reference with a detailed explanation on other access points that the user may want to consult.

general material designation: Abbreviated as GMD, this indicates to what category the material belongs, as far as format is concerned. Terms such as video recording, slide, and music, are used for the general material designation.

kit: Two different types of media combined as one unit, such as filmstrip and audiocassette, slide and booklet.

main entry: An access point used to identify a work. In the card environment, the main entry card contains the complete bibliographic information. For a book, the main entry is usually the author. If cataloged on a computer, all access points can be retrieved the same way; therefore, distinguishing between the main and added entries becomes unimportant.

microform: Microimages of materials produced on negative or positive films. The most common ones are microfilm (on reel) and microfiche (on 4 × 6-inch film) formats. An appropriate machine, the microfilm/fiche reader/printer, is necessary when using microform.

name-title reference: As an access point, a reference that includes both the name of the author and the title of the work.

physical description: A step in the cataloging process that involves describing the material physically, such as number of pages of the book, its size in centimeters, or the length of running time in the case of video and audio recordings.

realia: Artifact or object used in everyday living, such as a game or toy.

see also reference: A reference that directs the user from one term or name to other related terms or names.

see reference: A reference that directs the user from terms or names that are not used to terms and names that are used.

serial: A publication issued continuously at regular or irregular intervals with the intention of going on indefinitely. Journals and newspapers are examples of serials.

series: Separate publications that are related in topic or form. Besides having its own title, each series carries a collective title so that the whole set can be identified.

specific material designation: Located in the physical description area, this is a term indicating the specific type of material, such as sound cassette (as opposed to "sound recording" for general material designation); or microfilm reel (as opposed to "microform" for general material designation).

title proper: The chief part of the title, including the alternative title.

uniform title: In cataloging, a title that is chosen for a work published under various titles so that the work is easier for the user to retrieve.

INTRODUCTION

As explained in Chapter 3, descriptive cataloging is the first step in cataloging library materials. Descriptive cataloging involves describing the material first, then deciding the entries to complete the process. For the purpose of consistency, a reference book titled *Anglo-American Cataloguing Rules*, Second Edition, 1998 Revision (AACR2R) has been compiled and adopted by most libraries. This book includes the rules that every library must follow when performing descriptive cataloging. The rules were put together, and the book prepared, by the American Library Association, the Australian Committee on Cataloging, the British Library, the Canadian Committee on Cataloging, The Library Association (Great Britain), and the Library of Congress. Because of automation and the utilization of large central databases, more and more libraries have come to realize the advantage of conforming to the standards of the AACR2R rules. When a library joins as a member of a consortium, the library takes on the responsibility of cataloging materials according to AACR2R rules.

When the first edition of the *Anglo-American Cataloguing Rules* was published in 1967, most libraries began to catalog according to these rules. Although many changes were made in the second edition, published in 1978, it was not until 1981 that libraries totally adopted these new rules. The revised second edition was published in 1988, and again revised in 1998. This latest edition is the appropriate one to use and is the one discussed in this chapter. No fundamental or philosophical changes occurred in the 1998 edition.

Let us now look at AACR2R. As a library technician, you will be doing copy cataloging most of the time. You will be retrieving needed information from a computer database or printed sources in order to complete your cataloging tasks. In the occasional case when you cannot find the material already cataloged by someone else, you will need to perform original cataloging, which means that you must determine the information needed by technically reading the item you are cataloging. Technical reading means that you would look at the title page, the copyright page, and the table of contents and perhaps read a little bit into the chapters to determine the subject matter of the book and to get all the necessary information. You need to know exactly how the information is organized to transfer that information onto cards or to enter the information into the computer database, if your library has online cataloging. AACR2R spells out the rules for performing this task.

Besides the print edition, AACR2R is also available electronically on a CD-ROM titled *AACR2R-e,* and is included on another CD-ROM version, *Cataloger's Desktop,* together with some other publications that are used as references in cataloging. Although you need not memorize all of the rules in AACR2R, it is necessary to familiarize yourself with the basic rules, those which will be used daily when cataloging materials. For the more specialized rules, you need to know that they exist, and how to find the exact rule that applies to your case, either through the table of contents or the index.

In this chapter, the more basic rules listed in AACR2R are explained. Keep in mind that these are only some of the rules. For more specialized, less frequently used rules, AACR2R should be consulted directly. Any book on the subject of cataloging is not a substitute for AACR2R, and as a member of the cataloging staff, you must acquire a copy of the latest edition of AACR2R, which is the 1998 revision, as your sourcebook.

ANGLO-AMERICAN CATALOGUING RULES, *SECOND EDITION, 1998 REVISION*

Anglo-American Cataloguing Rules, Second Edition, 1998 Revision, or AACR2R for short, is divided into two parts.

Part I is titled *Description.* In the first thirteen chapters, detailed rules and formats are carefully explained for physically describing books, pamphlets, cartographic materials, manuscripts, music, sound recordings, motion pictures and video recordings, graphic materials, computer files, three-dimensional artifacts and realia, microforms, serials, and analysis.

Part II is titled *Headings, Uniform Titles, and References.* In Chapters 21 through 26, the following rules are found: choice of access point, headings for persons, geographic names, headings for corporate bodies, uniform titles, and references. A close study of each chapter will provide a complete explanation of each rule.

Chapters 14 through 20 are reserved to provide room for future expansion of the rules.

The appendixes section of the book covers (A) Capitalization, (B) Abbreviation, (C) Numerals, (D) Glossary. It is essential that ways to capitalize, to abbreviate, and to write out numbers, as detailed in Appendixes A, B, and C, are followed. Appendix D, the glossary, will assist beginners to more readily understand the text.

The index section in the back comes in handy when a specific question or problem surfaces. This section links the term being looked up to the rule number so that the appropriate rule can be located and applied.

In the following pages, rules are examined according to their original sequence in the AACR2R. Rule numbers are identified for reference purposes.

PART I. DESCRIPTION

In Chapter 1 of AACR2R, "General Rules for Description," the most basic rules are listed. Some of the more commonly used ones are outlined here.

1. Information for cataloging is to be taken from the "chief source of information." The chief sources for different types of materials are stated in the relevant chapters of AACR2R. If the chief source is lacking, data can be taken from any source. (Rules 1.0 A1, 1.0 A2)

2. The description is divided into the following areas. Each area may have more than one element.

Title and statement of responsibility

Edition

Material specific details

Publication, distribution, etc.

Physical description

Series

Note

Standard number and terms of availability

(Rule 1.0 B1)

For example, *Edition* is an area. *Publication, distribution, etc.* is another area, whereas *Publication* is an element.

3. These are the general guidelines for punctuation.

   a. Precede each area by a full stop, space, dash, space (. –), unless the area begins a new paragraph.
   b. Use square brackets ([ ]) to indicate that data are taken from outside the prescribed sources.
   c. Use an ellipsis (. . .) to indicate the omission of part of the element.
   d. General material designation (GMD) is always enclosed in its own brackets ([ ]). (Rule 1.0 C)

4. To suit the needs of libraries large and small, levels of detail in the description are established. A library should choose the level that is most appropriate to its size. Three levels are prescribed in Rule 1.0 D.

a. "First level" of description includes the following elements:

- Title proper / first statement of responsibility, if different from main entry heading in form or number or if there is no main entry heading
- Edition statement
- Material (or type of publication) specific details
- First publisher, etc., date of publication, etc.
- Extent of item
- Note(s)
- Standard number

b. "Second level" of description contains more details:

- Title proper [general material designation] = Parallel title : other title information / first statement of responsibility ; each subsequent statement of responsibility
- Edition statement / first statement of edition
- Material (or type of publication) specific details
- First place of publication, etc. : first publisher, etc., date of publication, etc.
- Extent of item : other physical details ; dimensions
- Title proper of series / statement of responsibility relating to series, ISSN of series ; numbering within subseries
- Note(s)
- Standard number

c. "Third level" of description includes all elements of the second level plus other information that is important for the library user.

Depending on the needs of each individual library, the appropriate level of description is chosen. Usually small libraries choose the first level of description, while medium libraries choose the second level. Only very large research libraries or special libraries practice the third level of description. Small libraries are required to do the second level of description if they belong to a consortium and have a shared database with other libraries.

AACR2R contains the details for each area of descriptive cataloging, starting with the title statement. Rules regulating all of the areas are summarized in the following sections.

Title and Statement of Responsibility Area (Rule 1.1)

Transcribe the title proper exactly as it is from the chief source of information in your material, except for punctuation and capitalization, which are prescribed in AACR2R in a separate place. (Rules for punctuation are listed in Rule 1.0 C, and rules for capitalization are stated in Appendix A.) If the collective title is shown, use the collective title.

The library may opt to use "general material designation," in brackets ([ ]), after the title. If used, terms must be taken from one of the two lists. List 1 refers to British libraries, and List 2 refers to libraries in the United States, Canada, and Australia.

List 1	List 2
braille	activity card
cartographic material	art original
computer file	art reproduction
graphic	braille
manuscript	chart
microform	computer file
motion picture	diorama
multimedia	filmstrip
music	flash card
object	game
sound recording	globe
text	kit
video recording	manuscript
	map
	microform
	microscope slide
	model
	motion picture
	music
	picture
	realia

slide

sound recording

technical drawing

text

toy

transparency

video recording

Example: Discovering the college library [video recording]

A parallel title follows the title proper, after an equals sign (=). Other title information also follows the title proper, separated by a colon (:).

Examples: The cat in the hat = Le chat au chapeau

Alice Walker : an annotated bibliography

Following the title and its related information, are statements of responsibility, separated by a slash (/). If there is more than one statement, a semicolon (;) is used for punctuation. In the statement of responsibility, the title, such as Dr., or PhD, is omitted, but the title of nobility, such as Baroness, is not.

Example: French legends, tales, and fairy stories / Retold by Barbara Leonie Picard ; illustrated by Joan Kiddell-Monroe.

Edition Area (Rule 1.2)

State the edition as indicated in abbreviated form, as instructed in Appendixes B and C in AACR2R.

Example: 3rd ed.

Any statement relating to such edition follows, separated by a slash (/), as described in the statement of responsibility section.

Example: 3rd ed. / revised and expanded by Eric J. Hunter.

Material (or Type of Publication) Specific Details Area (Rule 1.3)

Use only when describing cartographic materials, music, computer files, serial publications, and microforms.

> Examples: Scale ca. 1:50,000,000
>
> Vol. 1, no.1 (Jan./Feb. 1993)-

Publication, Distribution, etc. Area (Rule 1.4)

State the place of publisher or distributor first, then a colon (:), then the name of the publisher or distributor, in its shortest possible form. A comma follows the publisher, and after the comma, the date of publication, distribution, etc., and conclude the area with a period (.).

> Examples: New York: Dover, 1993.
>
> Guilford, CT: Annual Editions, 1993.

Physical Description Area (Rule 1.5)

State the number of pages of the book. If there is more than one volume, state the number of volumes. Use "ill." after a colon (:) to indicate illustrations. Add other descriptions such as maps after ill. and a comma. Give the height of the item in centimeters, and precede this with a semicolon (;). If accompanying materials exist, their description follows the height measurement and a plus sign (+).

> Examples: 568 p. : ill., maps ; 28 cm.
>
> 5v. : col. ports. ; 21 cm. + 1 answer book

Series Area (Rule 1.6)

The series statement is enclosed in parentheses [( )] following the physical description area. Included are series title, statement of responsibility, other information such as the International Standard Serial Number (abbreviated as ISSN), subseries, and the numbering within the series.

Examples: 568 p. : ill., maps ; 28 cm. (America in crisis)

216 p. : ports. ; 21 cm. (Graeco-Roman memoirs, ISSN 0306-9222 ; no. 62)

Note Area (Rule 1.7)

Start a new paragraph for note area. Notes are made for any additional important information that is not already included in areas described previously. Also, notes are made for the audience level, for the summary of the content, for the full or selective contents of the item, for library holding status, and for a "with" note that indicates this is only part of the item and a collective title is not available.

Examples: Play in 3 acts.

Library has v.1, 3-5, and 7 only.

Contents : Love and peril / the Marquis of Lorne – To be or not to be / Mrs. Alexander.

With: Candles at night / Alexander Napier.

Standard Number and Terms of Availability Area (Rule 1.8)

Start a new paragraph for the International Standard Book Number (ISBN). It is optional to give the price after the colon (:).

Example: ISBN 0-901212-04-0 (set) : $198.00

Supplementary Items (Rule 1.9)

Catalog the independent supplementary item separately from the main item as if they are two distinct items. For dependent supplementary items, either describe the minor supplementary item as accompanying material or as a note, or describe both items equally, one after another.

Examples: 5 v. : col. ports. ; 21 cm. + 1 answer book.

Accompanied by an answer book.

The music pact / Ron Van Der Meer and Michael Berkeley . . .

Music words : key definitions, key styles / . . .

Items Made Up of Several Types of Materials (Rule 1.10)

Items that are made up of two or more components are called kits. If one item is predominant, catalog the other part as a supplementary item, as described in Rule 1.9. If there is no predominant component, follow Rule 1.9, too, and describe both items as equal.

> Example: 1 filmstrip (52 fr.) : col. ; 35 mm.
> 1 sound cassette (45 min.) : analog, stereo.

Facsimiles, Photocopies, and Other Reproductions (Rule 1.11)

Describe facsimiles, photocopies, and other reproductions as stated in the chief source of information. Give data on original in the note area.

> Example: The human body / Editorial Board of Time-Life
> Books. – New York. Time-Life Books, 1995. Reprint of article from World Book Encyclopedia.

This concludes the AACR2R general rules of descriptive cataloging. The rest of this section examines and explains each chapter of AACR2R. For each chapter, only the most commonly used features are discussed. These rules do not include all possible situations encountered when doing descriptive cataloging. For items that contain specific features not covered in this text, AACR2R should be consulted in reference.

Chapters 2 through 12 detail the application of the rules that should be applied for different types of materials. Types of materials included are books, pamphlets, and printed materials; cartographic materials; manuscripts; music; sound recordings; motion pictures and video recordings; graphic materials; computer files; three-dimensional artifacts and realia; microforms; and serials. Each chapter explains the rules for the particular chosen medium. Let us discuss one medium at a time.

Chapter 2 of AACR2R specifies what rules are applied for cataloging books, pamphlets, and printed sheets. For these materials, the chief source of information is the title page. If there is no title page, the source from within the publication is used as a substitute. (Rule 2.0 B1)

Example: The wonderful adventures of Nils / by Selma Lager-
lof; translated from the Swedish by Velma Swan-
ston Howard; illustrated by H. Baumhauer. – Forge
Village, Mass. : Pantheon, 1911. – 539 p. ; 23 cm.
ISBN 0-123456-78-9

Chapter 3 explains special rules for cataloging cartographic materials, which, according to AACR2R, include "all materials that represent the whole or part of the earth or any celestial body. These include
two- and three-dimensional maps and plans; aeronautical, naviga-
tional, and celestial charts; atlases; globes; block diagrams; map sections; aerial photographs with a cartographic purpose; bird's-eye
views; etc." (1998, p. 94). The chief source of information for a
printed atlas is the title page. The chief source of information for
items other than an atlas is the cartographic item itself, or the case
that contains the item. If information is not available from the chief
source of information, it is taken from the accompanying printed materials. (Rules 3.0 A1, B2)

Example: Rand McNally cosmopolitan world atlas / cartography,
Michael W. Dobson ; design, Gordon Hartshorne. – New
ed. – Scale varies. – Chicago : Rand McNally,
1987. 1 atlas (viii, 288 p.) : col. : 27 cm.

Example: Rand McNally cosmopolitan world globe / design,
Gordon Hartshorne. – Scale 1:24,000. – Chicago:
Rand McNally, 1989. – 1 globe : col., wood ; 27
cm. in diam.

Chapter 4 lists the rules for cataloging manuscripts, including
manuscript collections. The rules used are the same as those designed
for printed materials. See example under Chapter 2 discussion.

Chapter 5 is for published music, which means music sheets, not
recorded music. The rules for cataloging music sheets are similar to
those of the printed materials. However, for music, special features
such as extent of the item and form of composition need to be recorded carefully. Also, the separately titled works in one item have to
be listed in the contents note. (Rules 5.5 B1 B2, 5.7 B1 B18)

Example: Charlie Brown's greatest hits [music] / music by
          Vince Guaraldi ; arranged by Lee Evans. – Milwau-
          kee, Wisconsin : Hal Leonard, 1984. – 3 scores
          (12 p.) : ill. ; 30 cm.
          Piano solos
          Three theme music from Vince Guaraldi for the
          Peanuts Television Specials
          Contents: Happiness theme – Linus and Lucy –
          Love will come.

Chapter 6 discusses sound recordings. The chief source of information for sound recordings is the item itself and its label. The physical description area needs special attention where specific material
designation is noted. Playing time is also recorded. Dimensions,
number of sound channels, and other pertinent information should
also be included. (Rules 6.0, 6.1, 6.5)

Example: John Denver's greatest hits [sound recording] / John
          Denver. – New York : RCA Records, 1973. – 1
          sound cassette (56 min.) : analog, stereo.

Chapter 7 covers motion pictures and video recordings, including
films, newscasts, programs, etc. The chief source of information for
motion pictures and video recordings is the title frames of the item, or
the label from its container, or the container itself. The gauge (width)
of motion pictures must be stated in millimeters, and videotapes in
inches, and the diameter of videodiscs must be given in inches.
(Rules 7.0, 7.5)

Example: Discovering the college library [video recording] /
          text by Marty Smith. – New York: Wilson, 1989. – 1
          videocassette (18 min.) : sd., col. ; ½ inch.

Chapter 8 of AACR2R describes any graphic materials, including
two-dimensional art originals, reproductions, charts, photographs, technical drawings, filmstrips, slides, etc. For general material designation
and for the physical description area, the following terms may be used
when cataloging graphic materials: activity card; art original; art repro-
duction; chart; filmstrip; flash card; picture; slide; technical drawing;

transparency. When an item has two different types of media, the term "kit" should be used for general material designation. The chief source of information for graphic materials is the item itself and the label on the container. (Rules 8.0, 8.5)

> Example: Dynasties in China [chart] / San Francisco : China Books & Periodicals, 1982. – 1 chart : b&w ; 27 x 21 cm.

Chapter 9 discusses the rules for describing the computer files that comprise data and programs, available both by direct and remote access. For computer files, the type of file should be indicated in the file characteristics area that follows the edition area. Terms applied here are computer data; computer program(s); computer data and program(s). The chief source of information for computer files is the title screen, or from the main menu, program statement, first display of information, the header to the file, etc. It is important to remember that in the note area, the nature and scope of the computer file and the system requirements should be stated. (Rules 9.0, 9.3, 9.5, 9.7)

> Example: Learn Microsoft Excel : intermediate and advanced [computer file] – Minneapolis, Minn. : Fast Start Learning, 1996. – Computer program – 1 computer optical disc : col. ; 4 ¾ in. + 1 user's guide.
>
> System requirements : IBM PC or compatible 486 or higher. 33 Mhz, 4MB RAM, CD-ROM drive, Windows 3.1/95, SVGA monitor with 640 x 480 resolution and 256 colors, Sound Blaster or compatible.

If a file is only available by remote access, the mode of access needs to be specified. In the note area, add a statement such as "Mode of access : Internet."

Chapter 10 is about three-dimensional artifacts and realia. Included in this category are models; dioramas; games; braille cassettes; sculptures; other three-dimensional artwork; exhibits; machines; clothing; microscope specimens; and other specimens mounted for viewing. The chief source of information is the object itself, together with any accompanying material or container. The dimensions of the object should be given in centimeters. If the object is in a container, the dimensions of the container should be given either after the dimensions of the object or as the only dimensions. (Rules 10.0, 10.5)

> Example: Monopoly [game] : Parker Brothers Real Estate Trading game. – Beverly, MA. : Parker Brothers, 1985. – 1 game: col. ; 50 x 25 x 3 cm.

Chapter 11 is about microforms. Microforms include microfilms, microfiches, micro-opaques, and aperture cards. The chief source of information for all is the title frame. For microfilm, its form, such as cartridge, cassette, or reel, should be added where appropriate. (Rules 11.0, 11.5)

> Example: The twentieth century [microform] : a pictorial history / photographs and text by Time-Life editors. – New York: Time-Life Books, 1999. – 2 microfilm reels: negative, ill. ; 35 mm.

Chapter 12 lists rules for cataloging serials. Serial is defined here as any title that is published continuously and intended to be published indefinitely. Serials are always dated or numbered in sequence. The interval may be regular, such as weekly magazines, daily newspapers, annuals, etc., or irregular, such as occasional papers and monographic series. The chief source of information for printed serials is the title page of the first issue of the serial. If the first issue is not available, the first available issue is used. Even though many smaller libraries do not catalog serials, library technicians still have to learn about serial cataloging because every library subscribes to many continuous publications, such as the almanac, annual guides, etc., that need to be cataloged. The nature of the serial is very different from all the media we have discussed so far. Because of its uniqueness, it is important for library technicians to gain knowledge on how to catalog serials. Some of the important points are summarized here. For nonprint serials, follow the rules for the particular medium. For example, for electronic journals, use the designated rules for computer files in Chapter 9. (Rules 12.1, 12.3, 12.5, 12.7)

1. When the title in the chief source of information is both in full and in acronym or initialism, choose the full form as title proper, and the other form as other title information. (Rule 21.1 B2)

   Example: The American journal of maternal / child nursing [serial] : MCN

2. A new description is made if the title changes. In other words, the new title is treated as a different title. (Rule 12.1 B8)

   Examples: American libraries – Vol. 1, no. 1 (Jan. 1970) – Continues : ALA Bulletin

   ALA Bulletin – Vol. 33-63 (Jan. 1939 – Dec. 1969) Continued by : American Libraries

3. Following the title, give the numeric designation first and then the chronological designation of the first issue of the serial. In the case of a completed serial, the designation of the final issue should also be noted. (Rules 12.3 C4 F1)

   Example: Nursing outlook – Vol.1, no.1 (Jan. 1953) – v. 40, no. 12 (Dec. 1992)

4. In many situations, notes need to be made for serials. The following are some of the more important ones:
   a. On the frequency of the serial, use terms such as annual, quarterly, irregular, etc.
   b. If titles of different issues vary, use "Title varies."
   c. If the publication is a translation of a previously published serial, use "Translation of" to start the note.
   d. If a serial continues another serial, add "Continues : " to start the note.
   e. If a serial is continued by another serial, use "Continued by:" to start the note.
   f. If a serial is a merger of two or more serials, use "Merger of : " or "Merger with : " to start the note.
   g. If a serial splits into two or more serials, use "Continues in part : " or "Split into : " to start the note.
   h. If a serial absorbs another serial, use "Absorbed : " to start the note.

    i. Make a note if a cumulative index is present.

    j. Make a note when there is something special included with the contents, such as exercises included for each chapter.

    k. Make a note of local holdings, the item's status, or re-strictions on its use.

Example: American libraries – Vol. 1, no. 1 (Jan. 1970) – 11 no. a year
                Indexed : Library Literature, Book Review Index
                Library has : Jan. 1980-
                Back issues in microfilm

Chapter 13 explains the rules for "Analysis." These detail how to describe a part or parts of an item, or how to indicate that the item in hand is part of a comprehensive work. As explained in the note area (Rule 1.7), parts of the work are usually listed as a content note in the note area. At other times though, a more comprehensive description may be necessary, and an "In" analytic entry is recommended. The word "In" should be italicized, underlined, or otherwise emphasized. (Rule 13.5)

Example: The Spring flowers
                In The best loved poems / New York: McGraw-Hill, 1999.

Whether working on cards or on the computer, the same AACR2R rules are to be followed. This means that even though the card and the online catalog look very different, cataloging is done the same way by following the same rules. It is easier to catalog online because there is a standard form, called the MARC format, that shows on the screen. The cataloger fills in the information for each separate line, under a different code number called a "field," corresponding to the chief source of information for the item, and then follows the prescribed way to enter the information. More on cataloging on the computer is explained in Chapter 8.

Both in the process of routine cataloging and when unusual problems and situations arise, it is necessary for the library technician to be familiar with all aspects of the AACR2R so that all the needed rules can be located efficiently.

After the description of the item is done according to the rules, the second step is to decide the main entry, that is, the first access point. Additional access points, called added entries, are also determined at this time. The rules on choosing the main entry, added entries, and how they should be stated properly are described in Part II of AACR2R, Chapters 21 through 26.

PART II. HEADINGS, UNIFORM TITLES, AND REFERENCES

Choice of Access Points (Chapter 21)

Some general rules about the choice of access points that every library technician should know are extracted from AACR2R and explained in this section:

1. Access points include main entry headings and added entry headings.

2. The chief source of information is used to determine the access points.

3. Enter a work by one or more persons under the principal author or the author named first. Make added entries for other names. (Rules 21.1 A2, 21.4 A1, 21.6 B1 B2 C1)

> Example: Introduction to Technical Services / Marty Bloomberg, G. Edward Evans
> Main entry: Bloomberg, Marty
> Added entry: Evans, G. Edward

4. Works such as internal policies, annual reports, directories, etc., of a corporate body are entered under the name of the corporate body. (Rules 21.1 B1 B2)

> Example: ALA handbook of organization and membership directory 1999-2000
> Main entry: American Library Association

5. Enter a work under its title if the following circumstances exist. (Rules 21.1 C1, 21.6 C2, 21.7 A1 B1)

- a. The personal authorship is unknown.
- b. The work is not the official publication of the corporate body.
- c. It is a collection of works by many authors. Make an added entry for the compiler or the editor in such a case.

Example: The virtual library : visions and realities / edited by
Laverna M. Saunders
Main entry: The virtual library
Added entry: Saunders, Laverna M.

- d. Editor instead of author is named.

Example: The nature and future of the catalog / edited by
Maurice J. Freeman and S. Michael Malinconico
Main entry: The nature and future of the catalog
Added entry: Malinconico, S. Michael

- e. If responsibility is shared and there are more than three compilers/editors, make an added entry for the first or principal compiler/editor.

Example: Texas county / Willie Nelson . . . [et al.]
Main entry: Texas county
Added entry: Nelson, Willie

6. If the title of a serial changes, make a separate main entry for each title. If any word of the title of a publication changes, consider it changed and enter it as a separate work. (Rules 21.2 A1 B1 C1)

7. Enter work that is modified or adapted from other work under the adapter or the modifier if the modification has substantially changed the nature of the work and if the work is paraphrased or re-written. (Rules 21.9, 21.10)

Example: Roget's thesaurus of English words and phrases /
completely revised and modernized by Robert A.

Dutch
Main entry: Dutch, Robert A.

8. Enter a work that consists of both text and illustrations under the name appropriate to the text. Make an added entry for the illustrator. (Rule 21.11)

Example: Insect / by Herbert S. Zim ; illustrated by James Gordon Irving
Main entry: Zim, Herbert S.
Added entry: Irving, James Gordon

9. Enter a work that is revised, updated, or enlarged:

a. Under the original author if the original author is named as being responsible for the work, make an added entry for the reviser. (Rule 21.12 A1)

b. If the original author is not considered to be responsible for the work for the newer edition, enter it under the reviser, or under the title, as appropriate. Make a name-title added entry for the original author. (Rule 21.12 B1)

10. Enter a translation under the heading appropriate to the original. Make an added entry for the translator. (Rule 21.14 A)

11. Enter a musical work under the composer. Make added entries for arranger, transcriber, writer, etc. If it is by various composers, enter it under title. Make an added entry for adaptor or arranger. (Rule 21.19 C1)

12. Enter a sound recording of one or more works by the author or the composer, whoever is appropriate. Make added entries for performers. In case of more than three performers, make an added entry for the first one only. (Rules 21.23 A1 B1)

13. If a sound recording contains works by different persons, enter it under the principal performer. If there are two or more performers, enter it under the first named and make added entries for the others. If four or more performers are listed, enter the recording under the title. (Rule 21.23 C1)

14. Added entries should be made as summarized in the following. (Rules 21.30 A1 B1 D1 E1 H1 J1 K1 K2 L1 M1)

   a. When the name of one person or one corporate body is used for the main entry, but two or three persons are responsible for the work, make added entries for the rest. If four or more names are involved, make an added entry for the one named first.
   b. When the editor or compiler is prominently named.
   c. When the corporate body or publisher has a substantial responsibility for the work.
   d. For any other name that would provide an important access point.
   e. For the illustrator.
   f. For the translator.
   g. For the heading of a series.
   h. For an analytical heading for a work contained within the item.
   i. For the title of every item entered under a personal or corporate entry.

15. Enter laws governing one jurisdiction under the heading for the jurisdiction and add a uniform title *[Laws, etc.]*. Make added entries for persons or corporate bodies responsible for compiling and issuing the laws. (Rule 21.31 B1)

   Example: General statutes of Connecticut : Revision of 1998
              Main entry: Connecticut
                 [Laws, etc.]
              Added entry: Connecticut. Legislative Commission-
              ers' Office

The previous information concludes Chapter 21 of AACR2R, "Choice of Access Point."

Originally, AACR2R was designed for the card environment, and, therefore, the rules show the distinction between the main and added entries, and how to choose both the main and added entries. In an automated environment, when cataloging is done on the computer using the MARC format, although main and added entries are entered into different fields, they are considered equally as access points and can be retrieved equally. The emphasis now is on ensuring that all the access points are entered in the MARC format.

Now that the main entry and the added entries are identified properly, rules for stating them correctly must be learned. These headings usually come in four different forms: persons, geographic names, corporate bodies, and uniform titles. Chapters 22 through 25 of AACR2R explain the rules for each of these four types of headings.

Headings for Persons (Chapter 22)

1. Use the name by which the person is commonly known. This may be a real name, pseudonym, title of nobility, nickname, initials, or other appellation. Treat a Roman numeral associated with a given name as part of the name. Determine the name from the chief source of information. (Rules 22.1 A B)

> Examples: Twain, Mark
> Theresa, Sister
> Seuss, Dr.
> Pope, John Paul II

2. If a person has changed his or her name, choose the latest name unless another name is better known. When name differs in fullness, the same rules apply. (Rules 22.2 C1, 22.3 A)

> Examples: Onassis, Jacqueline Kennedy
> Taylor, Elizabeth

3. If a person uses more than one name, use the name appearing in the work. Make references to connect the name. (Rule 22.2 B3)

> Example: Fast, Howard
> Reference: Fast, Howard
> *See also*
> Ericson, Walter
> Cunningham, E. V.

4. Enter a name containing a compound surname (consists of two or more proper names) under the element by which the person prefers to be entered. If this is unknown, check reference sources, such as a biographical dictionary, and follow the practice. (Rule 22.5 C2)

> Example: Lloyd George, David

5. Hyphenated names and other compound names are entered under the first element of the name. (Rules 22.5 C3 C4)

Examples: Day-Lewis, C.
           Johnson Smith, Geoffrey

6. If the name includes an article or preposition, or a combination of the two, enter the name under the element most commonly used in the person's language. (Rule 22.5 D1)

Examples: De la Mare, Walter (English name, under prefix)

Las Heras, Manuel Antonio (Spanish name with prefix, under prefix)

Casas, Bartolome de las (Spanish name with article and preposition, under the part following the prefix)

7. Enter a name that is a phrase or appellation in direct order. (Rules 22.11 A D)

Example: Poor Richard
           Author of *The moon river*

8. Additions to a name are made by adding the appropriate title or terms of address to the name in the vernacular. Add title to the name of a nobleman or noblewoman. Add *Saint* after the name of a Christian saint. Add a word or associated phrase when the name consists only of a surname. Add *Mrs.* to a married woman's name if she is only identified by her husband's name. For royalty, add a person's title. (Rules 22.12 A1, 22.13 A, 22.15 A)

Examples: Bismarck, Otto, Furst von

Gordon, Lord George
More, Sir Thomas, Saint
Smith, Mrs. Charles
Charles IV, King of France

9. Add a person's dates if the heading is otherwise identical to others. (Rule 22.17 A)

Example: Smith, Robert
           Smith, Robert, 1942-
           Smith, Robert, 1887-1953

10. Fuller names are added to the commonly used names if they are identical. (Rule 22.18 A)

Example: Johnson, A.H. (Allison Heartz)
Johnson, A.H. (Arthur Henry)

11. Consult rules 22.22 through 22.28 for names in certain languages other than English or those already mentioned.

Geographic Names (Chapter 23)

1. Use the English form of the place if there is one in general use. Use the name in the official language of the country if there is no English name for it. (Rules 23.2 A1 B1)

Examples: Austria (not Österreich)
Buenos Aires (no other English form in general use)

2. Add the name of a state, province, etc., to the name of a place for Australia, Canada, Malaysia, the United States, the former Soviet Union, and the former Yugoslavia. The names of states, provinces, territories, etc., of these countries are stated as is and need no addition. For other countries not listed here, add the name of the country in parentheses. (Rules 23.4 A1 B1 C1 C2)

Examples: Emeryville (California)
Connecticut
Prince Edward Island
Shangdong (China)

3. No addition is needed for the names of all parts of the British Isles. Add the name of the part in parentheses to the name of the place located there.

Examples: Northern Ireland
Wales
Bangor (Northern Ireland)
Powys (Wales)

4. To further identify the place, give the name of an appropriate smaller place before the name of a larger place in parentheses. (Rule 23.4 F2)

Example: Mohegan Park (Norwich, Conn.)

Headings for Corporate Bodies (Chapter 24)

1. Enter a corporate body directly under the name by which it is commonly identified. If the name consists of initials, omit or include a full stop according to the predominant usage. Do not leave space between the full stops. Do not leave space between the letters of an initialism written without full stops. (Rule 24.1 A)

Example: EDUCOM
           H.W. Wilson Foundation
           AFL-CIO

2. Use the conventional name of a government. (Rule 24.3 E1)

Examples: France (not République Française)
           Massachusetts (not Commonwealth
           of Massachusetts)

3. If the name alone does not convey the idea of a corporate body, add a general designation in English in parentheses. (Rule 24.4 B1)

Example: Apollo II (Spacecraft)

4. To distinguish between two of the same or similar names, a word or phrase in parentheses may be added to the headings. (Rules 24.4 C1-C7)

Examples: Democratic Party (Conn.)
           Asian Heritage Club (Stanford University)
           Pomona College (Claremont, Calif.)

5. Omit an initial article unless the heading is to be filed under the article. Omit the term or abbreviation indicating incorporation or ownership. (Rules 24.5 A1 C1-C4)

Examples: Library Association (not The Library Association)
Arizona (ship) (not U.S.S. Arizona)

6. For conferences, congresses, and meetings, add to the name the number, year, and place in parentheses. Separate these elements by a space, colon, space. (Rules 24.7 B2-B4)

Example: Off-Campus Quality Education Conference (13th :
1985 : Clearwater, Fla.)

7. If a subordinate of a corporate body is itself identifiable, enter under its own name. If it is not, make it a subheading. (Rules 24.12, 24.13)

Examples: Harvard Law School
Stanford University. Department of Economics
Yale University. Library

8. Enter the corporate body with a hierarchy under its name, with a subheading of its lowest element. For a government agency, enter under the government, and use the lowest element in the hierarchy as the subheading. Skip the names in the middle. (Rules 24.14, 24.19)

Example: American Library Association. Committee on Education for Library/Media Technicians
Hierarchy: American Library Association
Association of College and Research Libraries
Junior and Community College Library Services
Committee on Education for Library/ Media Tech-Technicians
United States. Office of Human Development Services
Hierarchy: United States
Department of Health, Education, and Welfare
Office of Human Development Services

9. For presidents and other heads of state, use the heading for the jurisdiction, followed by the title of the official, and add the years of the reign and the name of the person in a brief form. (Rule 24.20 B1)

Example: United States. President (1953-1961: Eisenhower)

10. For legislative bodies, enter under the name of the jurisdiction. Enter a committee or other subordinate unit as a subheading. Add the number and years in parentheses if available. (Rules 24.21 A-D)

Examples: United States. Congress. Joint Committee of the Library

United States. Congress (87th : 1961-1962). House of Representatives

United Kingdom. Parliament. House of Commons

Many more rules are listed in Chapter 24 of AACR2R. If in doubt, consult the book for the correct form for the corporate headings.

Uniform Titles (Chapter 25)

Some works are published under various titles. The case may be a different edition of the book or a translation of a work into a different language. It may also be a collection of different works. There is a need to bring all various titles together. AACR2R contains rules in Chapter 25 to take care of these situations by using what is called uniform title. Uniform title means one title is chosen for all variations of the titles, so that all titles are listed together in the catalog. Different editions of a work, even though titles vary, are not considered to be in this category. For different editions, new entries must be made.

1. Select one title as the uniform title if the work appears in various titles. Enclose the uniform title in square brackets before the title proper. For a title entry, it is optional whether or not to use the brackets. (Rule 25.2 A)

Examples: Dickens, Charles
[Pickwick papers]
The posthumous papers of the Pickwick Club
Arabian nights
One thousand and one nights

2. If a part of the work has a title of its own, use the title of the part as the uniform title. Make a "see" reference for the heading of the

whole work, and use the title of the part as a subheading. If the item has a consecutive number, use it as part of the subheading for the uniform title. If the item consists of three or more unnumbered parts, use the uniform title for the whole work followed by *Selections*. (Rules 25.6 A1 B1 B3)

Examples: Dickens, Charles
    [Hard times]
    See Dickens, Charles. Dickens' new stories
    Hard times
    See
    Dickens, Charles. Hard times
    Gibbon, Edward
    [History of the decline and fall of the Roman Empire. Selections]
    Homer
    [Iliad. Book 1-6]

3. Use the collective title *Works* for complete works of a person. Use the collective title *Selections* for items consisting of three or more works in various forms. Use the following collective titles for complete works of a person in a single form: *Correspondence, Essays, Novels, Plays, Poems, Prose Works, Short Stories, Speeches*. If the collection or selection of works is in a different language, add the language to the collective title in the brackets. (Rules 25.8 A, 25.9 A, 25.10 A, 25.11 A)

Examples: Maugham, W. Somerset
    [Works]
    Complete works

    Maugham, W. Somerset
    [Selections]
    Selected writings of Somerset W. Maugham

    Maugham, W. Somerset
    [Plays. Selections]
    Six great plays of Somerset W. Maugham

    Maugham, W. Somerset

> [Short stories. Spanish. Selections]
> En los mares del sur

4. For a complete or partial collection of laws and treaties, use [Laws, etc.] and [Treaties, etc.] after the jurisdiction. (Rules 25.15 A, 25.16 A1)

> Examples: Connecticut
>   [Laws, etc.]
>   General statutes of Connecticut . . .
>
>   United States
>   [Treaties, etc.]

5. For the Bible, enter a testament as a subheading of Bible. Then add the name of the language, followed by version and year. (Rules 25.18 A1 A10 A13)

> Example: Bible. N.T. English. Revised Standard. 1959

6. If a single selection of the Bible is commonly known by its own title, use that title as the uniform title. Make a "see" reference for the title in whole. (Rule 25.18 A7)

> Example: Ten commandments
> Reference: Bible. O.T. Exodus XX, 2-17
>   *See*
>   Ten commandments

7. Rules of uniform titles for musical titles are described together here:

> a.   Use the composer's original title. (Rule 25.27 A1)

> Example: Wagner, Richard
>   [Die meistersing von Nürnberg]
>   The mastersingers of Nürnberg

> b.   If the title includes the name of a type of composition, use the name of the type as the uniform title. (Rule 25.27 D)

> Example: Beethoven, Ludwig van
>   [Symphonies . . .]
>   Sinfonia eroica

c. For instrumental music, the uniform title may be certain standard chamber music combinations *(Trios, strings . . . ; Quartets, strings . . . ; Quartets, woodwinds . . . ; Quintets, winds . . . ; Trios, piano, strings . . . ; Quartets, piano, strings . . . ; Quintets, piano, strings . . .)*, individual instruments, or groups of instruments. (Rules 25.30 B2 B3 B4 B5)

Example: Debussy, Achille Claude
[Piano]
Suite bergamasque

These are just a few of the most commonly used rules. For any other situation not mentioned here, consult Chapter 25 of AACR2R. When cataloging musical works such as print music or recordings, the library technician must be familiar with all rules in AACR2R that govern musical works including Chapters 5, 6, and the appropriate section in Chapter 25.

References (Chapter 26)

This last chapter is about references. Depending on the needs, four types of references may be made. (Rules 26.1 B C D E)

1. *See* references. Make *see* references from a form that the library user may know to the form that has been chosen as the heading.

Examples: Clemens, Samuel Langhorne
See
Twain, Mark
International Business Machines
See
IBM

2. *See also* references. Make *see also* references from one heading to another related heading.

Examples: Pennsylvania. Department of Public Assistance
See also
Pennsylvania. Department of Welfare

Hibbert, Eleanor
See also
Carr, Philippa
Holt, Victoria
Kellow, Kathleen
(A person who writes under different names)

3. *Name-Title* references. Make a *see* or *see also* reference from a title that has been entered as part of the title in another entry.

Example: Tolkien, J.R.R.
Lord of the rings. 2, Two towers
See
Tolkien, J.R.R. Two towers

4. *Explanatory* references. Make an explanatory reference giving more explicit guidance when *see* and *see also* references are not adequate. The cataloger decides what explanatory references are needed and their wording.

Example: Conference . . .
Conference proceedings are entered under the name of the conference, or the title of the publication of the conference.

Rules for capitalization are explained in Appendix A, "Capitalization." In summary, capitalize headings in accordance with normal usage in the language. For the title, capitalize the first word. Do not capitalize words in a "general material designation." For any "area" in cataloging, capitalize the first word.

Appendix B, "Abbreviations," lists all the proper forms for abbreviating words, such as Conn. for Connecticut, bibl. for bibliography, Jan. for January, etc. Familiarity with the appendixes allows prompt and efficient answers to questions that arise.

The rules listed in AACR2R were strictly designed for a manual cataloging system and make no mention of, or accommodation for, libraries that catalog on computers. For example, although the automated catalog makes no distinction between main and added entries, these AACR2R rules still are applied to the computerized catalog. Both card and online cataloging must adhere to the requirements of AACR2R.

Also to be emphasized is that all print and nonprint materials should be cataloged according to the AACR2R rules. Materials of all formats should be listed together and intershelved so that the library can offer more efficient services to information seekers.

In this chapter, only the more frequently used rules have been discussed. In the event that the material you are cataloging does not fall into these categories, AACR2R should always be consulted.

REVIEW QUESTIONS

1. What is the purpose of AACR2R?
2. How is AACR2R organized?
3. List all the areas to be included in describing an item.
4. What accommodations are made in AACR2R to meet the different needs of large and small libraries?
5. What are the most common main entries and added entries?
6. What is a uniform title? What purposes does it serve?
7. Explain the different types of references.
8. Under what conditions is the main entry the title entry?

Chapter 5

Subject Headings

So far, you have studied the first two steps in cataloging: descriptive cataloging and choosing main and added entries. Now you are ready to take the third step: assigning subject headings.

In searching for information or doing research, a specific author or title often is not a primary consideration. Rather, most library users attempt to discover materials in a particular field by first checking under the appropriate subject. Thus, the task of assigning subject headings to materials takes on great importance. The purpose of assigning subject headings is to list all the materials on a given subject under a uniform term or phrase, so that in one search library users not only will be able to identify all the materials on a topic owned by the library, but also will find them side by side on the shelves. As with choosing main and added entries, discussed in Chapter 4, assigning subject headings also creates more access points for a work.

A subject heading may be a term; name of a person, group, or place; or a phrase. Obviously, many different terms can be applied to every subject. For example, the term movie can be called motion picture, cinema, film, and moving picture. If the choice and assigning system were left to the individual cataloger, or to the individual library, confusion would result. Similar to the use of rules for descriptive cataloging and for deciding main and added entries, reference books are used to maintain uniformity in assigning subject headings. *Library of Congress Subject Headings* is used by large libraries and network libraries, large or small. Some smaller public libraries and school libraries use the second reference book for this purpose, *Sears List of Subject Headings*.

In determining what subject headings to assign, the library technician should first determine what the material is about by, for example, in the case of a book, reading or looking at the book itself, its table of contents, or the blurb on the jacket. The title is often uninformative or misleading, and not a good source to use in trying to figure out the main topic of the material. Sometimes a more thorough examination is needed to completely understand what subjects the material covers. In such cases, more in-depth study of the material becomes necessary. If the subject still is not apparent, then references should be used, or a subject specialist should be consulted. Sometimes the language of the material is so specific, as in the case of foreign publications, that a language specialist should be consulted.

After the subject is determined, either *Library of Congress Subject Headings* or *Sears List of Subject Headings* is consulted to find the uniform term that then becomes the subject heading. This subject heading is either put on the bottom part of the catalog card, with an Arabic numbered sequence, or input onto the computer screen using the MARC format (see Chapter 8) 6XX fields designated for subject headings. For example, personal name headings are tagged 600, corporate name headings are tagged 610, topical headings are tagged 650, and geographical headings are tagged 651. Figure 5.1 shows the two subject headings printed on the bottom of the card: *Processing (Libraries)—Management* and *Library Administration.*

Figure 5.2 shows the same information on the computer screen where the subject headings are in the 650 field. The different kinds of

FIGURE 5.1. Subject Headings on a Card

Cargill, Jennifer S.

Z
688.5
.L48
1988

Library management and technical services : the changing role of technical services in Library organizations / Jennifer Cargill, editor.—New York : Haworth Press, c1988.
154 p. : ill. ; 23 cm.
Published as v. 9, no. 1 of the Journal of library administration.
Includes bibliographies.
ISBN 0-86656-779-8
1. Processing (Libraries)—Management. 2. Library administration. I. Cargill, Jennifer S.

18 NOV 88   17650291   MHGAdc   88-6824

FIGURE 5.2. Subject Headings on a Computer Screen

```
MHG —   FOR OTHER HOLDINGS, ENTER dh DEPRESS DISPLAY RECD SEND
OCLC:   17650291      Rec stat: p    Entrd:    880307      Used: 891026
Type:   a Bib lvl: m  Govt pub:      Lang:     eng Source:  Illus: a
Repr:   Enc lvl:      Conf pub: 0    Ctry:     nyu Dat tp: s  M/F/B: 00
Indx:   0 Mod rec:    Festschr: 0    Contr: b
Desc:   a Int lvl:    Dates: 1988,
     1  010           88—6824
     2  040           DLC  |c  DLC
     3  020           0866567798
     4  050 0         Z688.5  |b. .L48  1988
     5  082 0         025 / .02 |2  19
     6  049           MHGA
     7  245 00        Library management and technical services : |b the
changing role of technical services in library organizations / |c Jennifer S.
Cargill, editor.
     8  260 0         New York: |b Haworth Press, |c c1988.
     9  300           154 p. : |b ill. ; |c 23 cm.
    10  500           Published also as v. 9, number 1 of the Journal of
library administration.
    11  504           Includes bibliographies.
 →  12  650 0         Processing (Libraries) |x Management.
 →  13  650 0         Library administration.
    14  700 10        Cargill, Jennifer S.
```

subdivisions are marked by the MARC format subfields, preceded by a delimiter, $ or ≠ or |. For example, |× for topical subdivisions, |v for form subdivisions, | y for period subdivisions, and |z for geographical subdivisions, explained in detail with examples later on in the chapter.

Because of the expense of printing, filing, and maintaining extra cards, it is recommended that subject headings assigned to an item in a card system library be limited to three or four. For automated libraries, the cataloging staff is encouraged to assign as many subject headings as deemed necessary.

If subject headings need to be more specific, subject subdivisions are added to the subject headings, separated by a dash (–). Subject subdivisions may be the description of physical forms, such as *Maps* or *Dictionaries;* they may be topical, such as *History* or *Study and Teaching;* they may be geographical, such as *Paris (France)* or *United*

States; they may be chronological, such as *History–1945-1953.* In any case, the directions outlined in *Library of Congress Subject Headings* or *Sears List of Subject Headings* must be followed.

Besides the terms used for assigning subject headings, both *Library of Congress Subject Headings* and *Sears List of Subject Headings* offer different kinds of references to guide users to more related materials. References given include "USE" references that direct users to the chosen terms, and "SA" references that list more related terms so that users can get added relevant information. Also, for the users' convenience, broader and narrower topics (BT and NT) are listed.

This chapter discusses these two reference tools in detail. Pay special attention to the one that your library uses.

TERMINOLOGY

authority file: A set of the established authoritative forms of headings, according to the Library of Congress or the local library, used in cataloging for inputting bibliographic records and for making references to and from the headings. A file of all the established subject headings is called the *subject authority file.* A file of all the established names is called the *name authority file.* A file of all the established series tiles is called the *series authority file.* A card catalog can serve as the authority file. The computerized catalog usually has built-in authority files generated by the Library of Congress. Any bibliographic utility or commercial library service company may provide an authority file.

BT: Stands for broader topic(s). Broader topics are listed so that the user can find more related information on the subject.

free-floating subdivision: A list of subdivisions and how they may be applied in conjunction with subject headings.

NT: Stands for narrower topic(s). Terms listed under NT are more specific and may add in-depth materials for the researcher.

pattern heading: For each category of subject heading, a standardized set of subdivisions is developed. To save space, these subdivisions are printed under only one subject heading belonging to the category, and this subject heading is called a pattern heading.

RT: Lists the related topic(s). These are other subject headings that in some manner relate to the topic.

SA: This is a "see also" reference that refers the user to an entire group of additional headings.

scope note: A note under the subject heading to explain and clarify how the term should be used, and noting what is included and what is not included when the term is used.

subject subdivision: Words or phrases following a subject heading after a dash that make the subject more specific. The subdivision may be topical, a physical form, geographical, or chronological.

UF: Means "used for," which is opposite from USE. The term before UF is the chosen heading. The one listed afterward is not to be used.

USE: Indicates that the term listed is not to be used because it is not a uniform term that has been chosen as a legitimate term. The term listed after "USE" should be used instead.

LIBRARY OF CONGRESS SUBJECT HEADINGS

Now in its twenty-second edition, published in 1999, *Library of Congress Subject Headings,* or LCSH for short, has a more than one-hundred-year history of development. It is an accumulation of subject headings established by the Library of Congress since 1898 and is the most comprehensive list of subject headings in the world, used by thousands of libraries as well as commercial indexes. It provides an alphabetical list of all subject headings, cross-references, and subdivisions. Since thousands of headings with subdivisions are added to LCSH each year, the Library of Congress has published the following aids for catalogers to use in conjunction with *Library of Congress Subject Headings.* They are published on a regular basis and contain information on policies and actions of the Library of Congress regarding changes of subject headings. The first one is *Subject Cataloging Manual: Subject Headings*, Fifth Edition (1996). This is a very useful how-to guide for assigning subject headings and subdivisions in a consistent way, as is practiced by the catalogers in the Library of Congress. A separate update, published annually, includes new headings and provides cross-references for changes. *Cataloging Service Bulletin* contains subject headings of current interest that have been recently changed since the latest edition of the *Manual*. Another supplement published annually for the LCSH is titled *Free-Floating Subdivisions: An Alphabetical Index*, Eleventh Edition (1999). This volume lists the subdivisions that do not appear in the subject authority file in the main volume but may be attached to certain subject headings. Another publication is *LC Period Subdivisions Under Names of Places*, Fifth Edition (1994), which lists subject headings for place names, with date subdivisions listed chronologically.

A new edition of *Library of Congress Subject Headings* is published every year. The latest edition is in five volumes, usually referred to by library staff as "the big red books." To keep information current, a weekly update is posted in an electronic version titled *LC Subject Headings Weekly Lists*, which is made available through the Library of Congress Cataloging Policy and Support Office Web page, <lcweb. loc.gov/catdir/cpso/>. It provides up-to-date information on new additions, deletions, or changes of subject headings, which have been examined, discussed, and approved by the editorial board composed of the Library of Congress staff and representatives from the library world. Another way to keep up with the changes and updates of all the Library of Congress publications, including LCSH, is through subscription to the electronic publication *LC Cataloging Newsline* (LCCN). This title is published irregularly and is available free of charge in electronic form only through the LCCN Web page <www.lcweb.loc.gov/catdir >. Also, regular reading of the aforementioned *Cataloging Service Bulletin* enables catalogers to keep up with the changes.

Besides the print volumes, a microfiche edition is produced quarterly for subscribers. LCSH can be found in *Classification Plus*, a full-text, Windows-based CD-ROM product that includes the *Library of Congress Classification Schedules* and *Library of Congress Subject Headings*. *Classification Plus* is available from the Library of Congress as an annual subscription with quarterly issues. Some commercial library automation companies also provide subject headings on CD-ROM to libraries. Selected new Library of Congress subject headings have been linked to the Dewey Decimal Classification numbers and are available on the Dewey home page <www.oclc.org/fp/l>.

New headings are added and outdated terms are deleted on a regular basis. Headings to be used are listed in **boldfaced** type. Subject headings come in different forms and lengths. The headings may be just one word, such as **Art,** or may be two words, usually an adjective and a noun, such as **Art students**. Headings may appear in inverted form, such as **Art, American.** A heading may be followed by the legend *(May Subd Geog)*. For broad topics, subdivisions apply, such as **Art—Study and teaching.** Sometimes headings may include conjunctions and prepositional phrases, such as **Art and literature** and **Art in literature,** or they may be a person's name, a corporate body's name, or a geographical name. Although these examples demonstrate varied forms and structures, a careful and patient search will match what is being cataloged and, thus, the proper procedure will come into effect. According to LCSH, 36 percent of headings are followed by Library of Congress Classification numbers. Students are warned, however, that the class numbers are offered as a guide only, and, therefore, verification in the *Library of Congress Classification Schedules* is necessary.

Some general principles for assigning and constructing subject headings according to *Library of Congress Subject Headings* and *Subject Cataloging Manual: Subject Headings* include the following:

1. Assign one or more subject headings that best summarize the contents of the work. The heading must represent at least 20 percent of the work.
2. Assign as many headings as needed.
3. Assign headings that are as specific as the topics of the work. If precise headings are not possible, assign broader or more general headings.
4. Assign a more general heading if the heading includes two or three related topics in the work and the heading represents no other topics.

5. If the work is about two or three topics and the general heading includes more than these topics, assign the two or three specific headings.
6. If there are four subtopics, assign four specific headings instead of a general heading.
7. Use names of persons, families, corporate bodies, projects, places, etc., as subject headings. Check any name authority file for the proper spelling, order, etc.
8. When more than one subject heading is chosen, the first one should represent the main topic of the item.

To assign appropriate subject headings to materials is most important because most library users retrieve materials by subject. If the assigned subject headings are too broad, too many not-so-useful materials may be retrieved. On the contrary, if they are too narrow, not enough information can be retrieved. Many more rules exist for assigning subject headings than those cited here. Consult *Subject Heading Manual: Subject Headings* for additional information. One has to keep in mind that Library of Congress policy and practice on assigning subject headings was designed for the card catalog environment, and, therefore, minimal redundancy was required. In an automated environment, however, it is convenient to increase the number of subject headings and to build in any degree of redundancy, with the intention of representing the subject completely.

In *Library of Congress Subject Headings,* under each heading, the conditions of use are explained through five kinds of references.

1. *Scope note.* Scope note follows the heading immediately to ensure consistency of subject usage. Scope note clarifies the range of subject matter to which a heading is applied. It also brings attention to the distinctions between related headings.

Example: **Art** *(May Subd Geog)*

Here are entered general works on the visual arts. Works on the arts in general, including visual arts, literature, and the performing arts, are entered under **Arts.**

2. *USE* and *UF references*. The USE reference is made from an unauthorized term to an authorized term. USE follows the unauthorized heading, but precedes the authorized one. UF, which stands for "used for," is the opposite of USE and comes between the authorized heading and the unauthorized one. USE is commonly called a *see* reference, which is made for synonyms, variant spellings, and different heading constructions.

> Examples: **Art** *(May Subd Geog)*
> UFArt, Western
> Arts, Visual
> Visual arts
> Western arts
> Visual arts
>
> USE **Art**

3. *BT* and *NT*. BT is for broader topics and NT is for narrower topics. By using headings under BT, one can find materials more general than the one originally checked. In the same sense, NT suggests more specific headings.

> Example: **Art** *(May Subd Geog)*
> BT **Arts**
> NT **Children's Art**
> **Collectors and Collecting**
> **Drawing**
> **Posters**

4. *RT*. RT stands for related topics. It gives related terms expressing any useful relationship other than BT and/or NT. It links two headings that are associated in some manner.

> Example: **Art**
> RT **Aesthetics**

5. *SA*. SA is a *see also* reference. This is a general blanket reference made to an entire group of headings.

Example: **Art**
           SA headings of the type [topic] in art, e.g., Christian
           saints in art

Complex topics in many cases are listed with subdivisions. Subdivisions exist to express different concepts or perspectives of the topic. A subdivision follows a dash (–) under the heading. If two subdivisions are used, the second one follows the first one with two dashes (– –).

Example: **Art**
           –Private collections
           – –Italy

Using this example, the complete subject heading should be:

Art–Private collections–Italy

Or, if it is on the computer screen, it would look like this:

650 0 **Art** $x **Private collections** $z **Italy**

Some subdivisions are listed and can be found in the text of LCSH, while others may be assigned according to the rules specified in the *Library of Congress Subject Headings Manual: Subject Headings*. There are four categories of subdivisions.

1. *Topical* subdivisions are used to limit the concept of the topic. Following is how it will appear on a card and in MARC format on the computer screen:

Example: **Art–Marketing**
           650 0 **Art** $x **Marketing**

2. *Form* subdivisions represent what a work is as far as form is concerned, not what the work is about.

Example: **Art–Slides**
           650 0 **Art** $v **Slides**

3. *Chronological* subdivisions limit a heading to a specific period of time.

Example: **Art–16th century**
650 0 **Art $y 16th century**

4. *Geographic* subdivisions indicate the geographic location. Instruction for the use of geographic subdivisions appears after the heading in a scope note or is indicated by *(May Subd Geog),* which means may be subdivided geographically. Rules for listing geographic subdivisions are outlined in the *Library of Congress Subject Headings Manual: Subject Headings.* Generally, the established name of the country is used if the material is about a country. If the geographic entity is a region or a city, then the name of the country precedes the name of the region or city. The exception is for the United States, Great Britain, Canada, and the former Soviet Union, where the name of the country is not used.

Example: **Art–France**
650 0 **Art $z France**
Art–New York
650 0 **Art $z New York**

To avoid repeating subdivisions under all possible headings of the same category, such as sports, diseases, or musical instruments, the *Library of Congress Subject Headings* established tables referred to as "Table of Pattern Headings" and "Free-Floating Subdivisions" appropriate for the same category. These pattern headings appear in a table in LCSH.

Examples:	*Category*	*Pattern Heading*
	Musical instrument	Piano
	Diseases	Cancer

These examples mean that when assigning subdivisions for **Saxophone,** the pattern for **Piano** is used since both belong to the musical instrument category. When trying to find subdivisions for **Heart Attack,** one has to look under **Cancer** and use the same subdivisions.

Standardized subdivisions used under different pattern headings are called free-floating subdivisions. A Library of Congress publication titled *Free-Floating Subdivisions: An Alphabetical Index* should be consulted for what subdivisions are available and how they may be used.

Example: *Subdivision*   *Category*
Methods–Group instruction   Musical instruments

In this case, the subdivision and the subject heading are linked together as follows:

Saxophone–Methods–Group instruction
650 0 **Saxophone $x Methods $x Group instruction**

Some categories of headings are omitted from the LCSH list. They include headings that appear in the following name authority files: *Name Authorities Cumulative Microform Edition, NACO Participants' Manual*, or other name authority lists provided by commercial automation vendors. For more information on name authority files, see the following Web site: <http://lcweb.loc.gov/cds/name_aut.tml>: headings that are created by need but are not those in the authority record, and certain music headings. The *Manual* should be consulted when adding headings in these categories to ensure the practice is consistent with the rules.

"The Annotated Card Program: AC Subject Headings" section in the LCSH lists subject headings used exclusively for the juvenile collection, tailored to the needs of children and young adults. Some headings are standard subject headings taken from the main volume of LCSH, some are modified subject headings from the same source, and some are new headings established to use only when assigning subject headings to juvenile literature with special subdivisions. These special subdivisions include Biography, Collections, Fiction, Guides, Habits and behavior, Illustrations, Pictorial works, or Wit and humor.

The primary intention of "AC Subject Headings" is to provide a liberal extension of Library of Congress subject headings so that a more appropriate and in-depth subject treatment can be achieved for juvenile titles.

In applying AC subject headings, some practices differ from LC subject headings:

1. Subdivision with the word "juvenile" is not used.
2. The use of the subdivision "United States" and the qualifying term "American" is restricted to topics that are international in scope, such as **Art** and **Music.**
3. For classes of persons, geographic subdivision is omitted, such as **Athletes** and **Actors.**
4. Subject headings are assigned to fiction.

There are many other rules to follow when assigning AC headings. Read the "Introduction" section in the *Library of Congress Subject Headings* for details.

Figure 5.3 shows the top of a page in *Library of Congress Subject Headings,* Twenty-second Edition, and Figure 5.4 shows a sample page from *Annotated Card Program: AC Subject Headings.*

SEARS LIST OF SUBJECT HEADINGS

Sears List of Subject Headings was developed to meet the needs of small to medium-sized libraries. The latest edition, the sixteenth, was published in 1997 by the H. W. Wilson Company. Some useful suggestions and pointers on how to use this book to find the appropriate subject headings are stated in a section titled "Principles of the *Sears List of Subject Headings.*" Generally speaking, the following are the most basic:

1. Assign a specific and direct heading. For example, use **Penguins,** not Birds. Use **Orange,** not Citrus. But if the material contains information about oranges, lemons, and other kinds of citrus fruits, use **Citrus Fruit.**
2. Apply common usage. For example, use **Labor,** not Labour. Use **Elevators,** not Lifts.
3. Use terms that are uniform. For example, use **Porcelain,** not China, or Chinaware.
4. Form headings such as **Essays, Poetry, Fiction, Hymns,** and **Songs** are used under **Collection** only, not under individual authors.

FIGURE 5.3. Sample from *Library of Congress Subject Headings*

Library stack management
    USE Stack management (Libraries)
Library stacks
    USE Library shelving
Library staff manuals
    USE Libraries—Staff manuals
Library stamps
    [Z689]
Library statistics
    [Z669.8 (Statistical methods)]
    [Z711.3 (Reference and use)]
  Here are entered works on the compilation and study of statistics of libraries, or collections of general library statistics. Other collections of statistics are entered under Libraries—[place], types of libraries, or names of individual libraries, with subdivision Statistics, e.g. Libraries—Wisconsin—Statistics; Academic libraries—Statistics.
    UF   Libraries—Statistical methods
            Libraries—Statistics
            Library science—Statistical methods
    BT   Statistics
    RT   Bibliometrics
            Books—Statistics
    NT   Library use studies
Library students, Interchange of
    USE   Librarian, exchange programs
Library supervisors, School
    USE   School library supervisors
Library supplies
    USE   Library fittings and supplies
Library surveys *(May Subd Geog)*
    BT   surveys
Library technicians *(May Subd Geog)*
    UF   Library assistants
            Library paraprofessionals
            Paraprofessionals in libraries
    BT   Library employees
Library technology specialists
    USE   Computer specialists in libraries
Library telephone reference services
    USE   Telephone reference services (Libraries)
Library teletype systems
    USE   Teletype in libraries
Library trustees *(May Subd Geog)*
    UF   Libraries—Trustees
            [Former heading]
            Library boards
            Trustees, Library
    BT   Trusts and trustees
    NT   Minority library trustees
            Public library trustees
            Women library trustees
Library use skills
    USE   Library research
Library use studies *(May Subd Geog)*
    [Z711.3]
    UF   Libraries—Use studies
    BT   Information services—Use studies
            Library statistics
    NT   Academic libraries—Use studies
            Agricultural libraries—Use studies
            Catalogs, Card—Use studies
            Catalogs, Subject—Use studies
            Children's libraries—Use studies
            Government publications—Use studies
            High school libraries—Use Studies
            Indexes—Use studies

Libraries—Circulation analysis
Library catalogs—Use studies
Library catalogs on microfilm—Use
Periodicals—Use studies
Public libraries—Use studies
Research libraries—Use studies
Social science literature—Use studies
Special libraries—Use studies
Subject headings—Use studies
Technical college libraries—Use studies
Young adults' libraries—Use studies
Library user orientation
    USE   Library orientation
Library volunteers
    USE   Volunteer workers in libraries
Library Week
    USE   National Library Week
Libration of the moon
    USE   Moon—Libration
Libration points
    USE   Lagrangian points
Librettists *(May Subd Geog)*
    [ML2110]
    UF   Opera—Librettists
    BT   Dramatists
            Libretto
            Musicians
Libretto
  Here are entered works on the history and criticism of the libretto and on libretto writing. Collections of miscellaneous librettos are entered under the heading Librettos.
    BT   Dramatic music
    NT   Librettists
Librettos
    [ML48-ML49]
  Here are entered collections of miscellaneous librettos. Collections of librettos limited to a specific form are entered under that form, e.g. Operas—Librettos; Oratorios—Librettos. Works on the history and criticism of the libretto and on libretto writing are entered under the heading Libretto.
    BT   Dramatic music
LIBRIS (Information retrieval system)
    UF   Library Information System
    BT   Information storage and retrieval systems
Libtako (Burkina Faso)
    USE   Liptako (Burkina Faso)
Libtrot family
            USE Liptrap family
Liburnia *(May Subd Geog)*
    [QL527.D44]
    BT   Delphacidae
Liby (Saltvik, Finland)
    BT   Farms—Finland
Libya
— **Antiquities**
    NT   Agora (Cyrene)
            Cyrene (Extinct city)
            Leptis Magna (Extinct city)
            Qasr as-Sahābī (Libya)
            Sabratha (Extinct city)
            Sanctuary of Demeter and Persephone (Cyrene)
            Temple of Hercules (Sabratha)
— **Antiquities Roman**
— **Description and travel**
    UF   Libya—Description and travel—1981- *[Former heading]*
— —1981-
    USE   Libya—Description and travel
—Foreign relations *(May Subd Geog)*

FIGURE 5.4. Sample Page from Annotated Card Program: AC Subject Headings

UF Folklore, Altai

Amaryllis (Hippeastrum)
UF Hippeastrum

American chameleon
USE Anoles

American [Danish, English, etc.] poetry
Here are entered single poems or collections of poetry by individual American [Danish, English, etc.] authors.
Collections of poetry by several authors of the same nationality are entered under American Danish, English, etc. poetry—Collections.

American [Danish, English, etc.] poetry Collections

American drama (Comedy)
USE Humorous plays

American wit and humor
USE Wit and humor

Amusements
NT Treasure hunts

Anastasia, Grand Duchess, daughter of Nicholas II, Emperor of Russia, 1901-1918
UF Anastasiia Nikolaevna, Grand Duchess, daughter of Nicholas II, Emperor of Russia, 1901-1918

Anastasiia Nikolaevna, Grand Duchess, daughter of Nicholas II, Emperor of Russia, 1901-1918
USE Anastasia, Grand Duchess, daughter of Nicholas II, Emperor of Russia, 1901-1918

Androids
USE Robots

Angel fish
USE Angelfish

Angelfish *(May Subd Geog)*
UF Angel fish
Freshwater angelfishes
Scalare

Animal behavior
USE Animals–Habits and behavior

Animal distribution
UF Biogeography

Grammar of *Sears* subject headings includes single and compound nouns, adjectives with nouns, and phrase headings. Similar to the *Library of Congress Subject Headings, Sears* also has four kinds of subdivisions: physical form, special aspect or topic, chronology, and place.

Symbols used in the sixteenth edition of *Sears List of Subject Headings* are the same as those used in the *Library of Congress Subject Headings*. These include UF, SA, BT, NT, and RT, and they maintain the same meanings as explained earlier in this chapter when discussing the *Library of Congress Subject Headings*. The *Sears* also uses *(May Sub Geog)* to indicate that the heading may be subdivided geographically, and a general scope note explaining the heading may immediately follow the heading.

Libraries usually have local rules on the maximum number of subject headings assigned to an item, which are typed on the bottom of cards following an Arabic number. In automated libraries, any number of the desired subject headings are keyed into the 6XX fields for subject headings. *See* and *See also* cards need to be typed for libraries with manual cataloging. In most automated libraries, the subject authority files are already established by the vendor or the bibliographic utility, and, therefore, the cataloging staff need not manually determine all individual references.

Other features in the *Sears List of Subject Headings* with which the library technician needs to be familiar include "Headings to Be Added by the Cataloger," "Key Headings," and "List of Commonly Used Subdivisions," among others. *Sears List of Subject Headings* is also in electronic format, to which libraries may subscribe through the Wilson Database Licensing Service.

After most of the *Sears* headings, one can find one or two suggested Dewey numbers for the topic taken from the *Abridged Dewey Decimal Classification and Relative Index*.

Figure 5.5 shows the top of a page from *Sears List of Subject Headings*. A comparison with Figure 5.3, the sample from the *Library of Congress Subject Headings*, shows that they basically follow the same principles, only the LC headings are more numerous, specific, and detailed. Because of record sharing with other automated libraries, many consortia now require their members to use the *Library of Congress Subject Headings* and not *Sears*.

FIGURE 5.5. Sample from *Sears List of Subject Headings*

Library science—*Continued*
NT   **Cataloging**
      Classification—Books
      Comparative librarianship
      Library surveys
      Library technical processes
RT   **Bibliography**
      Library services
Library science—Study and teaching
USE   **Library education**
Library services 025.5

Use for materials on services offered by libraries to patrons. General materials on the knowledge and skill necessary for the organization and administration of libraries are entered under **Library science.**

UF   Reader services (Libraries)
      [Former heading]
SA   Libraries and specific types of users or specific activities for which services are provided, e.g. **Libraries and the elderly;** to be added as needed
BT   **Libraries**
NT   **Bibliographic instruction**
      Libraries and African Americans
      Libraries and labor
      Libraries and students
      Libraries and the elderly
      Library circulation
      Library extension
      Reference services (Libraries)
      Young adults' library services
RT   **Library science**
Library services to African Americans
USE   **Libraries and African Americans**
Library services to children
USE   **Children's libraries**
Library services to labor
USE   **Libraries and labor**
Library services to teenagers
USE   **Young adults' library services**
Library services to the elderly
USE   **Libraries and the elderly**
Library services to young adults
USE   **Young adults' library services**
Library skills
USE   **Bibliographic instruction**

Library supplies
USE   **Libraries—Equipment and supplies**
Library surveys 020
BT   **Library science**
Library systems
USE   **Libraries—Centralization**
      Library information networks
Library technical processes 025
Use for materials on the activities and processes concerned with the acquisition, organization, and preparation of library materials for use.
UF   Centralized processing (Libraries)
      Libraries—Technical services
      Library processing
      Processing (Libraries)
      Technical services (Libraries)
BT   **Libraries**
      Library science
NT   **Cataloging**
      Classification—Books
      Libraries—Acquisitions
      Libraries—Collection development
Library technicians 020.92
UF   Library assistants
      Library clerks
      Paraprofessional librarians
BT   **Librarians**
      Paraprofessionals
Library trustees
USE   **Libraries—Trustees**
Library unions
USE   **Librarians' unions**
Library user orientation
USE   **Bibliographic instruction**
Librettos 780; 780.26
Use for collections of miscellaneous librettos and for materials on the history and criticism of librettos and on writing librettos. Individual librettos and collections of librettos of a specific type are entered under the specific type of libretto.
SA   types of librettos, e.g.
      Opera librettos; to be added as needed
NT   **Opera librettos**
Lie detectors and detection 363.2
UF   Polygraph
BT   **Criminal investigation**

BT = Broader Term   NT = Narrower Term   RT = Related Term   SA = See Also   UF = Used For

Source: Sears List of Subject Headings, 15th edition, p. 404. Copyright © 1994 by the H. W. Wilson Company. Material reproduced with permission of the publisher.

REVIEW QUESTIONS

1. What is the purpose of assigning subject headings to materials?
2. What are the two reference tools used to assign subject headings?
3. List the procedures for determining subject headings to be used.
4. Explain all the reference symbols used in the *Library of Congress Subject Headings*.
5. When is subdivision for a subject heading necessary?
6. List all the different types of subdivisions.

Chapter 6

Classification

Before the classification system is discussed, let us review the steps involved in the cataloging process so that it is clear where the task of classification belongs. Recall the first step of cataloging, called *descriptive cataloging,* is performed according to the rules in AACR2R. Within this step, we have learned how to physically describe the material at hand, and how to determine the main entry and the added entries. Then we continued to the second step, called *subject cataloging.* The task of subject cataloging includes two parts: assigning subject headings and assigning classification numbers. Chapter 5 explained the process of assigning subject headings to materials. Now we are ready for the discussion on classification.

In supermarkets or in stores, a classification system arranges all the same and related merchandise together on the shelves so that it is easier for the customers to find what they want. In a library, this objective is even more important. First of all, the classification system is important because the classification number, combined with the author number, forms a unique notation that becomes the location symbol for the material. This "address" makes it convenient for library users to find the materials on the shelves. In addition, the system is designed so that materials of the same subject are grouped together on the shelves, with related subjects nearby for easy browsing. Moreover, with the unique location symbol labeled on the spine, the library staff can easily shelve the material in its proper place after each use. Whether a library technician works in the cataloging department or in public services, a thorough knowledge of a library's classification scheme is essential to better serve library users. A classification system provides a powerful tool to extract information not only from the library catalogs but from the global information network as well.

Different classification systems have been designed for adoption by libraries, but only two have been universally adopted: the Dewey Decimal Classification system and the Library of Congress Classification system. The library technician needs to understand these classification systems in order to assign numbers correctly to the materials, and also to help users locate materials. A library usually uses one system exclusively.

Generally speaking, most academic libraries and special libraries use the Library of Congress Classification system. Most public libraries and school libraries use the Dewey Decimal Classification system. Many academic libraries switched from the Dewey system to the Library of Congress system in the late 1960s and early 1970s, and, therefore, in these libraries, both systems may be in operation. Some older books will still carry Dewey numbers, while the newer acquisitions will have the Library of Congress numbers.

The preliminary processes for assigning a classification number to the material are the same as assigning subject headings, as discussed in Chapter 5. First of all, one has to know exactly to what subject field the material belongs by carefully examining the table of contents, the preface, or even the text of the material. For audiovisual materials, if the labels or the containers do not supply enough information for making a decision, the library technician should listen to or view the contents. In some extremely difficult cases, one may have to consult subject specialists.

Whereas several subject headings can be assigned to a single item, the classification number has a limit of one. If an item deals with several subjects, the dominant one should be used. If the dominant subject cannot be decided, the first mentioned should be used. If the item deals with more than three subjects, a broader, inclusive subject should be chosen. After the subject matter is determined, the classification schedule used by the library is consulted, and the number or notation that appropriately describes the item's subject matter is chosen.

In this chapter, both the Dewey Decimal Classification system and the Library of Congress Classification system will be examined and explained in detail, together with examples.

TERMINOLOGY

author number: *See* BOOK NUMBER.

book number: A letter and number combination assigned to a particular book representing the author. Combined with the classification number, it forms a unique symbol. This ensures that books on the same subject will be shelved side by side, alphabetically by authors' last names. Book number is also called author number or cutter number.

call number: The location, or the address, of an item on the shelves. The call number is a combination of the classification number and the book number. Sometimes other notations such as volume number, notation representing the title, and other symbols are assigned by the cataloger so that each item has a unique notation. This notation is called a call number—the number to call for the material.

cutter number: *See also* BOOK NUMBER. Book numbers are derived from a reference book titled *Cutter-Sanborn Three-Figure Author Table* and are therefore called the Cutter number. Two other versions of this book are titled *C. A. Cutter's Three-Figure Author Table* and *Cutter's Two-Figure Author Table*.

notation: A number or number and letter combination that is prescribed in the classification scheme.

DEWEY DECIMAL CLASSIFICATION SYSTEM

The Dewey Decimal Classification system is the most widely used classification system in the world. This reference tool, called DDC in short, now in its twenty-first edition, published in 1996, contains four volumes. As stated in the introduction to the Dewey Classification, this system is used by over 200,000 libraries in 135 countries and has been translated into over thirty foreign languages. In the United States, 95 percent of all public and school libraries, 25 percent of all college and university libraries, and 20 percent of special libraries use this system. Various bibliographic utilities and online services use the DDC system as their retrieval tool.

A detailed discussion of the four volumes of DDC 21 follows. Volume one is titled *Introduction. Tables.* The use of this reference book and the principles of this classification scheme are explained in the introduction section. Some main concepts are especially important to remember. The basic classes are arranged by traditional academic disciplines, not by subject. A given subject may be found in more than one place, depending on the different aspects of the subject. For example, a book on the topic of *Internet* may be given different classification numbers, according to DDC 21. The number 004.678 is used if the book is an interdisciplinary work about the Internet and contains information about computer hardware. If the book does not contain much information about computer science, but the emphasis is on search and retrieval, number 025.04 would be used. For works that describe Internet information resources devoted to specific disciplines and subjects, 025.06 is assigned, and, finally, works on Internet access providers and works on economic and public policy issues concerning the Internet are classified as 384.33.

The entire world of knowledge is divided into ten main classes. The ten classes are further divided into ten divisions, and each division is divided into ten sections. All classification numbers are three-digit numbers, such as 000, 130, 269, etc. In the three-digit number, the first digit indicates the main class, the second digit indicates the division, and the third digit indicates the section. Following the three-digit number is a decimal point after which may be additional digits that more specifically define the subject.

For example, for number 025.0 indicates the main class *Generalities.* The middle number 2 indicates division *Library and information sciences,* and number 5 indicates the section *Operations of libraries, archives, information centers.* To be more specific, the number 025.43 as the classification number for a book indicates that the book is about library classification systems.

Seven tables used to limit the aspect of a subject are included in volume one of DDC 21. For examples, from Table 1, "Standard Subdivisions," number 092 indicates persons under a main topic. From Table 2, "Geographic Areas, Periods, Persons," number 73 means United States under a main topic. The same kind of arrangement can be found in Table 3A, "Individual Authors"; Table 3B, "More Than One Author"; Table 3C, "Notation to Be Added"; Table 4, "Subdivisions of

Individual Languages and Language Families"; Table 5, "Racial, Ethnic, National Groups"; Table 6, "Languages"; and Table 7, "Groups of Persons." Please remember that numbers in the tables are never used alone; they are always added to the classification number. For example, in number 810.8, 810 is American literature in English, the broad discipline; while .8 represents miscellaneous writings of individual authors, which is derived from Table 3.

Last in volume one are lists that compare the previous edition with the new edition: "Relocations and Reductions"; "Comparative and Equivalence Tables"; and "Reused Numbers." A new book needs to be classified with a new number, of course. Depending on the size of the library, its budget, and its staffing situation, some libraries may decide that it is not cost-efficient to reclassify older books, and only new acquisitions will be classified according to the twenty-first edition.

Volumes two and three are titled *The Schedules,* which are referred to when determining the appropriate number for the material being classified. The schedules are divided into ten categories of three-digit numbers, as described earlier. Three summaries that are partially reproduced here give outlines of the organization.

Figure 6.1 shows the "First Summary," which divides the knowledge in the world into ten categories corresponding to discipline fields of study. In performing the task of classifying, one first must decide which category describes the topic most closely. For example, for a foreign language dictionary, 400 is the appropriate class to choose.

Figure 6.2, the "Second Summary," shows the divisions applied to the second number, called the hundred divisions. This means that for each of the classes in the "First Summary," ten divisions are formed to further break down the discipline to make it more specific. For example, for a book on education, it is first decided that the field *Education* belongs to the class of social science 300. Further examining the "Second Summary" will lead to number 370 for Education.

Figure 6.3, the "Third Summary" for class number one, shows the divisions applied to the third digit, called the thousand sections. For instance, a book on *Higher Education* will have the number 378, meaning that the broad class is social science, the division is education, and the section is higher education.

Following the summaries are the main schedules in volumes two and three. Class 000 through class 500 are included in volume two, and

FIGURE 6.1. "First Summary"

The Ten Main Classes

000	Generalities
100	Philosophy & psychology
200	Religion
300	Social sciences
400	Language
500	Natural sciences & mathematics
600	Technology (Applied sciences)
700	The arts. Fine and decorative arts
800	Literature & rhetoric
900	Geography & history

class 600 through class 900 in volume three. To facilitate the job of assigning Dewey numbers, library technicians should familiarize themselves with the schedules and the pattern of each category.

By checking the main schedules, one will find that more details may be added following a decimal point. For example, for the classification number 378.25:

300	denotes social science
370	denotes education
378	denotes higher education
378.2	denotes academic degrees
378.25	denotes honorary degrees

Also, numbers from the tables in volume one of DDC 21 may be added. For example, for the classification number 510.711:

500	denotes natural sciences and mathematics
510	denotes mathematics
510.7	denotes education, research, related topics (from Table 1)
510.71	denotes schools and courses (from Table 1)
510.711	denotes higher education (from Table 1)

FIGURE 6.2. "Second Summary"*: The Hundred Divisions

000 Generalities
010 Bibliography
020 Library & information sciences
030 General encyclopedic works
040
050 General serial publications
060 General organizations & museology
070 News media, journalism, publishing
080 General collections
090 Manuscripts & rare books

100 Philosophy & psychology
110 Metaphysics
120 Epistemology, causation, humankind
130 Paranormal phenomena
140 Specific philosophical schools
150 Psychology
160 Logic
170 Ethics (Moral philosophy)
180 Ancient, medieval, Oriental philosophy
190 Modern western philosophy

200 Religion
210 Philosophy & theory of religion
220 Bible
230 Christianity   Christian theology
240 Christian moral & devotional theology
250 Christian orders & local church
260 Social & ecclesiastical theology
270 History of Christianity & Christian church
280 Christian denominations & sects
290 Comparative religion & other religions

300 Social sciences
310 Collections of general statistics
320 Political science
330 Economics
340 Law
350 Public administration & military science
360 Social problems & services; associations
370 Education
380 Commerce, communications, transportation
390 Customs, etiquette, folklore

400 Language
410 Linguistics
420 English & Old English
430 Germanic languages   German
440 Romance languages   French
450 Italian, Romanian, Rhaeto-Romanic
460 Spanish & Portuguese languages
470 Italic languages   Latin
480 Hellenic languages   Classical Greek
490 Other languages

500 Natural sciences & mathematics
510 Mathematics
520 Astronomy & allied sciences
530 Physics
540 Chemistry & allied sciences
550 Earth sciences
560 Paleontology   Paleozoology
570 Life sciences   Biology
580 Plants
590 Animals

600 Technology (Applied sciences)
610 Medical sciences   Medicine
620 Engineering & allied operations
630 Agriculture & related technologies
640 Home economics & family living
650 Management & auxiliary services
660 Chemical engineering
670 Manufacturing
680 Manufacture for specific uses
690 Buildings

700 The arts   Fine & decorative arts
710 Civic & landscape art
720 Architecture
730 Plastic arts   Sculpture
740 Drawing & decorative arts
750 Painting & paintings
760 Graphic arts   Printmaking & prints
770 Photography & photographs
780 Music
790 Recreational & performing arts

800 Literature & rhetoric
810 American literature in English
820 English & Old English literatures
830 Literatures of Germanic languages
840 Literatures of Romance languages
850 Italian, Romanian, Rhaeto-Romanic
860 Spanish & Portuguese literatures
870 Italic literatures   Latin
880 Hellenic literatures   Classical Greek
890 Literatures of other languages

900 Geography & history
910 Geography & travel
920 Biography, genealogy, insignia
930 History of ancient world to ca. 499
940 General history of Europe
950 General history of Asia   Far East
960 General history of Africa
970 General history of North America
980 General history of South America
990 General history of other areas

Source: Reproduced from the Dewey Decimal Classification, published in 1989, by permission of OCLC Forest Press, a division of OCLC Online Computer Library Center, owner of copyright.

*Consult schedules for complete and exact headings.

FIGURE 6.3. "Third Summary"*: The Thousand Divisions

000	**Generalities**
001	Knowledge
002	The book
003	Systems
004	Data processing Computer science
005	Computer programming, programs, data
006	Special computer methods
007	
008	
009	

010	**Bibliography**
011	Bibliographies
012	Bibliographies of individuals
013	Of works by specific classes of authors
014	Of anonymous and pseudonymous works
015	Of works from specific places
016	Of works on specific subjects
017	General subject catalogs
018	Catalogs arranged by author, date, etc.
019	Dictionary catalogs

020	**Library & information sciences**
021	Library relationships
022	Administration of the physical plant
023	Personnel administration
024	
025	Library operations
026	Libraries for specific subjects
027	General libraries
028	Reading & use of other information media
029	

030	**General encyclopedic works**
031	American English-language
032	In English
033	In other Germanic languages
034	In French, Provençal, Catalan
035	In Italian, Romanian, Rhaeto-Romanic
036	In Spanish & Portuguese
037	In Slavic languages
038	In Scandinavian languages
039	In other languages

040	
041	
042	
043	
044	
045	
046	
047	
048	
049	

050	**General serial publications**
051	American English-language
052	In English
053	In other Germanic languages
054	In French, Provençal, Catalan
055	In Italian, Romanian, Rhaeto-Romanic
056	In Spanish & Portuguese
057	In Slavic languages
058	In Scandinavian languages
059	In other languages

060	**General or organizations & museology**
061	In North America
062	In British Isles In England
063	In central Europe In Germany
064	In France & Monaco
065	In Italy & adjacent territories
066	In Iberian Peninsula & adjacent islands
067	In eastern Europe In Russia
068	In other geographic areas
069	Museology (Museum science)

070	**New media, journalism, publishing**
071	Journalism & newspapers in North America
072	In British Isles In England
073	In central Europe In Germany
074	In France & Monaco
075	In Italy and adjacent territories
076	In Iberian Peninsula and adjacent islands
077	In eastern Europe In Russia
078	In Scandinavia
079	In other geographic areas

080	**General collections**
081	American English-language
082	General collections in English
083	In other Germanic languages
084	In French, Provençal, Catalan
085	In Italian, Romanian, Rhaeto-Romanic
086	In Spanish & Portuguese
087	In Slavic languages
088	In Scandinavian languages
089	In Italic, Hellenic, other languages

090	**Manuscripts & rare books**
091	Manuscripts
092	Block books
093	Incunabula
094	Printed books
095	Books notable for bindings
096	Books notable for illustrations
097	Books notable for ownership or origin
098	Prohibited works, forgeries, hoaxes
099	Books notable for format

*Consult schedules for complete and exact headings.

Volume four is titled *Relative Index. Manual.* The *Relative Index* is the first place the cataloger will go to find the approximate schedule of details. For example, if the material is about the library's role in higher education, looking in this volume under the heading *Education*, one would find 021.24 for this subject. If using volume two directly and checking under *Education* in the social science category, one would be hard-pressed to find the appropriate number. Reference notes will guide the DDC 21 users to the correct number. The second part, *Manual*, explains the policies and practices of how the schedules are designed, updated, and changed. Also, notes on the tables offer help and suggestions so that students of cataloging can get a clear idea of how to perform the challenging task of assigning Dewey numbers to the material at hand.

Beginner library technicians should adhere to the following procedures in classifying materials. First, determine the discipline. Then the Dewey summaries are checked to determine the appropriate three-digit number to match the discipline. After that, the *Schedules*, either volume two or volume three, are consulted to find the exact decimal numbers to use after the initial three-digit number. Depending on the topic and the explanation following the number, a further decimal number from the table may have to be added to complete the Dewey number. If unsure of the appropriate number for the discipline, one can always check the *Relative Index* of volume four, which provides a guide to the exact number in the main schedules. Pay special attention to the copious notes that are included in the schedules, giving explanations and directions on how to build the number, how to use the tables, and other important instructions.

A shorter version of DDC 21, titled *Abridged Edition 13: Dewey Decimal Classification,* is simpler and is designed for small libraries with small collections. The thirteenth edition, published in 1997, is based on the contents of the DDC 21.

DDC 21 is also available in a CD-ROM format called *Dewey for Windows.* Automated libraries may use this version, which offers full-text indexing, full-text schedules, a personal notepad, and Library of Congress subject headings linked to Dewey numbers. The search for Dewey numbers is made faster and more efficient when using *Dewey for Windows* because the cataloger can search by words, phrases, numbers, index terms, and Boolean combinations. It even offers automatic cuttering, which saves the cataloger one step in the cataloging process.

Between editions, the Dewey Classification system is kept up to date by an annual publication from the Forest Press titled *Dewey Decimal Classification: Additions, Notes and Decisions,* or DC& for short. The most current new and changed entries are posted on the first day of each month on World Wide Web site <oclc.org/oclc/fp/ddc/newnchng.htm>. Forest Press's home page, <oclc.org/oclc/fp>, also offers news and enhancements related to the Dewey schedules.

BOOK NUMBER FOR THE DDC

The classification number alone is not sufficient to make a unique identification of each item in the library. In cataloging, an extra step is taken to assign a book number, also referred to as the author number, or the Cutter number. The combination of the classification number and the book number gives every item in the library a unique notation that places the item in a specific location. Frequently, the notation includes prefixes, which are placed above the classification numbers. The most common ones are *Ref* or *Reference* for reference materials, *J* or *Juv* for juvenile materials, *E* or *Easy* for easy-reading collections, etc. This combined notation is called the call number, which is what we see on the spine labels of library materials.

The book number starts with a decimal point followed by a letter, which is the first letter of the author's last name. If there is no author, the first letter of the title is often used. Following the letter is a numerical value that represents the alphabetical order of the author's last name. Because many items may bear the same Dewey classification number, the book number allows items on the same subject to be shelved alphabetically by author's last name.

The three versions of the reference tool for assigning the book number are *C. A. Cutter's Three-Figure Author Table, Cutter-Sanborn Three-Figure Author Table*, and a simple *Cutter's Two-Figure Author Table*. Because of the growth in library collections, which created a need for an extended Cutter number, in 1996, OCLC published the *Four-Figure Cutter Tables,* based on the existing three-figure schemes. See Figure 6.4 for a sample page reproduced from *C. A. Cutter's Three-Figure Author Ta-*

ble. See Figure 6.5 for a sample page from the *Four-Figure Cutter Tables,* which can be found in OCLC Special Report at <oclc.org/oclc/research/publications/review96/cutter.htm>.

For example, the call number for the book *Personal Power* by Arleen Labella is 650.14.L112. The Dewey number for this book is 650.14, indicating the subject of the book. To be more exact, 650 is for the subject *Management and auxiliary services,* and .14 is one of the subdivisions for 650 indicating the more specific topic *Success in obtaining jobs and promotion.* The author number .L112 gives this book an exact location on the shelves, between a book on the same topic by Lab (.L111) and another book on the same topic by Labi (.L113).

More illustrations on how to decide on the author number are described here. For author Samuel Baker, the author number is .B177. Margaret Armstrong's author number is .Ar58. For Edward Smith, the author number is .Sm55. When the author's name falls between two numbers, the first one is used. For example, for author Tidd, .T438 is the appropriate author number. For author Range, .R163 is the right choice. The number next to the author's name is chosen according to instructions printed in the Cutter book, such as using the first two letters of an author's name if the first letter is a vowel.

Occasionally two or more call numbers will be the same, making it difficult to decide on their absolute location on the shelves. In this case,

FIGURE 6.4. Sample from *C. A. Cutter's Three-Figure Author Table*

L	1	M							
La	11	Ma	Lamo	19	Mc	K	Lann	28	Mai
Lab	111	Maag	Lamond	191	"	Kai	Lanne	281	Maie
Labau	112	Maas	Lamor	192	"	Kau	Lanner	282	Mailf
Labi	113	Mab	Lamorr	193	"	Ke	Lannes	283	Mailles
Labori	114	Mabill	Lamou	194	"	Kee	Lanno	284	Maimo
Labro	115	Mabl	Lampa	195	"	Kell	Lano	285	Mainf
Lacane	116	Macaf	Lampe	196	"	Ken	Lanot	286	Mair
Lachau	117	Macal	Lamper	197	"	Kenep	Lanou	287	Maiso
Lacho	118	Macan	Lampl	198	"	Kener	Lanoye	288	Maiu
Lacom	119	Macas	Lampr	199	"	Keners	Lanq	289	Maju

FIGURE 6.5. Sample from *Four-Figure Cutter Tables*

Am185 Ambrose K	Am3547 American Ho	Am46 Amhe
Am186 Ambrose Ste	Am3548 American Host	Am47 Amherst
Am187 Ambrosi	Am3549 American Indian R	Am48 Amherst J
Am188 Ambrosini M	Am3551 American Institute Of	Am49 Amhu
Am189 Ambroz	Architects D	Am51 Ami
Am19 Ambu	Am3552 American Institute Of	Am52 Amie
Am21 Amc	Certified Public	Am53 Amien
Am22 Amcc	Accountants Co	Am54 Amin
Am23 Amcl	Am3553 American Institute Of Con	Am55 Amint
Am24 Amco	Am3554 American Institute Of Pr	Am56 Amir
Am25 Amcr	Am3555 American Iss	Am57 Amis
Am26 Amcu	Am3556 American Juv	Am58 Aml
Am27 Amd	Am3557 American Legion Au	Am59 Amlo
Am28 Amdo	Am3558 American Library	Am61 Amm
Am29 Amdr	Association Boo	Am62 Amme
Am31 Ame	Am3559 American Library	Am63 Ammi
Am32 Amel	Association Res	Am64 Ammo
Am33 Amelo	Am3561 American Lux	Am65 Amn
Am34 Amen	Am3562 American Management	AM66 Amo
Am3511 Amer	Association A	Am67 Amon
Am3512 America Hi	Am3563 American Medic	Am68 Amor
Am3513 America Ph	Am3564 American Meter	Am69 Amos
Am3514 American Academy Of Op	Am3565 American Nat	Am71 Amp
Am3515 American Academy Of Ps	Am3566 American National Standard	Am72 Amper
Am3516 American Antique	For Ph	Am73 Ampf
Am3517 American Ass	Am3567 American National Standards	Am74 Amph
Am3518 American Association For Th	I	Am75 Ampu
	Am3568 American New	Am76 Amr
Am3519 American Association Of Co	Am3569 American Oi	Am77 Amrap
Am3521 American Association Of N	Am3571 American Pet	
Am3522 American Association of St	Am3572 American Pho	
	Am3573 American Polity	
	Am3574 American Psychiatric G	
	Am3575 American Public We	
	Am3576 American Re	
	Am3577 American Revolution C	
	Am3578 American School Of Co	
	Am3579 American Society For Co	
	Am3581 American Society	
	For S Am3582 American	
	Society	
	For Testing	
	Materials C	

another notation is added to the call number, referred to as the work mark. The work mark is decided by the cataloger and may involve adding the publication date to the call numbers, which has become a common practice. For example:

650.14	650.14	650.14
.L112	.L112	.L112
1976	1985	1992

The work mark may involve attaching the first letter, and sometimes the first two or even three letters, of the title to the call numbers. For example:

650.14	650.14	650.14
.L112Al	.L112Am	.L112C

A special instruction printed in *C. A. Cutter's Three-Figure Author Table* shows that under some circumstances, more than one initial letter is used. The instruction reads, "Use one letter for words beginning with consonants (except S), two for words beginning with vowels and with S, three for words beginning with Sc. Letters I, O, U, and X need usually only one figure. Ii, Iw, Ix, Iy, Oo, Uo, Uq, Uu, Ss, and Sx can generally be used without figures." For example, for Smith, the author number is .Sm69; for Schmid, the author number is .Sch52.

Some libraries do not use the Cutter *tables* to figure out the author numbers. Instead, authors' last names are used as the author number, or in some cases, the first three letters of an author's last name are used.

In MARC format, the Dewey call number is always entered in the field with a 082 tag number. If the number is locally assigned, it is entered after tag 092.

LIBRARY OF CONGRESS CLASSIFICATION SYSTEM

The Library of Congress Classification system is the second most widely used system in the United States, used by most academic libraries and special libraries. Many of these libraries used the Dewey system originally but changed to the Library of Congress system in the late 1960s and early 1970s.

The Library of Congress Classification (LCC) system divides knowledge into twenty-one broad categories, using a letter to represent each subject field. The letters I, O, W, X, and Y are excluded. See Figure 6.6 for the outline of the schedules.

The Library of Congress Classification system is used much the same way as the Dewey system, except that letters are substituted for numbers to denote the particular subject field. To provide for expansion, double or triple letters, when necessary, are used for subclasses. The letter notation along with a numerical value completes the LC classification number. On some occasions, another letter-number combination is necessary to describe precisely the subject matter. For example, the LC classification number for the book *Training for Non-Trainers* is HF5549.5.T7. The following breakdown shows what each part or number represents:

H	Social science
HF	Commerce
HF5549	Personnel management
HF5549.5	Employment management. By topic, A-Z
HF5549.5.T7	Training of employee

Note that .T7 is part of the classification number, not an author number or book number.

Each of the LC class schedules is published separately in pamphlet format. The individual subject schemes are independently created by subject specialists in each field and therefore do not fall into a consistent pattern. Since this system was devised originally by the Library of Congress to organize its own collection, some schedules are more detailed than others. When needed, each schedule can be expanded easily by adding numbers, decimals, and letters to the main class, and, in fact, this is done regularly by the Library of Congress. Each schedule has a different publication date, and the new revisions are printed at different intervals.

Each schedule includes a synopsis first, then the schedule, and finally an index. Because there is no general overall index, it is the responsibility of the classifier to choose the appropriate schedule for the subject matter in hand. To help find the most appropriate schedule, the *LC Classification Outline* provides some general help. The

FIGURE 6.6. Library of Congress Classification Schedules Outline

A	General Works. 5th ed. (1998)
B-BJ	Philosophy. Psychology (1996)
BL,BM,BP,BQ	Religion: Religions. Hinduism, Judaism, Islam, Buddhism (1984)
BR-BV	Religion: Christianity, Bible (1987)
BX	Religion: Christian Denominations (1985)
C	Auxiliary Sciences of History (1996)
D-DJ	History (General), History of Europe, Part 1, 3rd ed. (1990)
DJK-DK	History of Eastern Europe: General, Soviet Union, Poland (1987)
DL-DR	History of Europe, Part 2, 3rd ed. (1990)
DS-DX	History of Asia, Africa, Australia, New Zealand, etc. (1998)
E-F	History: America (1995)
G	Geography. Maps. Anthropology. Recreation. 4th ed. (1976)
GE	Environmental Science (1976)
H	Social Sciences (1997)
J	Political Science (1997)
K	Law (General) 1998 edition
K Tables	K Tables: Form Division Tables for Law. 1999 edition
KD	Law of the United Kingdom and Ireland (1998)
KDZ, KG-KH	Law of the Americas, Latin America, and the West Indies (1984)
KE	Law of Canada (1998)
KF	Law of the United States (1999)
KJ-KKZ	Law of Europe (1989)
KJV-KJW	Law of France (1999)
KK-KKC	Law of Germany (1982)
KL-KWX	Law of Asia and Eurasia, Africa, Pacific Area and Antarctica (1993)
KZ	Law of Nations (1998)
L	Education (1998)
M	Music and Books on Music (1998)
N	Fine Arts (1996)
P-PA	Philology and Linguistics (General). Greek Languages and Literature Latin Language and Literature (1997)
P-PZ Tables	Language and Literature Tables (1998)
PB-PH	Modern European Languages (1999)
PJ-PK	Oriental Philology and Literature, Indo-Iranian Philology and Literature. (1988)
PL-PM	Languages of Eastern Asia, Africa, Oceania; Hyperborean, Indian, and Artificial Languages (1988)
P-PM	Supplement: Index to Languages and Dialects (1991)
PN	Literature (General) (1997)
PR-PS, PZ	English and American Literature, Juvenile Belles Lettres (1998)
PQ	French, Italian, Spanish, and Portuguese Literature (1998)
PT, Part 1	German Literature (1989)
PT, Part 2	Dutch and Scandinavian Literature (1992)
Q	Science (1996)
R	Medicine (1999)
S	Agriculture (1996)
T	Technology (1999)
U-V	Military Science. Naval Science (1996)
Z	Bibliography and Library Science (1995)

Library of Congress Subject Headings, which was discussed in Chapter 5, may be used as an index because suggested LC numbers are printed after almost all subject headings. The Library of Congress also publishes *Subject Cataloging Manual: Classification* as a practical, time-saving, how-to manual that helps catalogers assign LC classification numbers. Some principles of classification, similar to those discussed in Chapter 5, include: classifying a work by its specific subject, using the most specific number available; assigning numbers according to instructions printed in the schedules; and classifying a work with a broader subject if the work deals with several subjects.

The LC schedules are kept up to date through the quarterly publication *Library of Congress Classification—Additions and Changes.* A CD-ROM format of the LC classification, called *Classification Plus,* produced by the Library of Congress, is also available. This is convenient and efficient because it allows a search by key words, classification numbers, proximity options, Boolean operations, etc. However, all schedules are not yet automated, so the latest CD-ROM version contains only twenty-seven schedules out of forty-six. Additional schedules will be added as they become available throughout the year. Another available electronic format is the *Super LCCS CD,* produced by Gale Research Company.

BOOK NUMBER FOR THE LCCS

The purpose of cuttering is to create a unique call number by composing a logical and orderly subarrangement within a class. The main element of the LC call number consists of one to three capital letters followed by a one- to four-digit number with up to three decimal places, for example, HF5549.5. A topical Cutter number is added to the classification number if an extraordinarily precise identification is required, for example, HF5549.5.T7. The .T7 stands for *training* in this example; it is not the author number.

The Library of Congress has devised its own method of assigning the book number, also called author number or Cutter number. It resembles the Cutter number from the *Cutter-Sanborn table,* but is less complicated because all the details are included and coded by the LC classification number itself, as described previously. Figure 6.7 shows the cutter numbers devised by the Library of Congress.

FIGURE 6.7. Cutter Numbers Devised by the Library of Congress

CUTTER TABLE

Library of Congress book numbers are composed of the initial letter of the main heading followed by Arabic numerals representing the succeeding letters on the following basis:

1. After initial vowels

for the 2nd letter:	b	d	l,m	n	p	r	s,t	u-y
use number:	2	3	4	5	6	7	8	9

2. After initial letter S

for the 2nd letter:	a	ch	e	h,i	m-p	t	u
use number:	2	3	4	5	6	7-8	9

3. After initial letters Qu

for the 3rd letter:	a	e	i	o	r	y
use number:	3	4	5	6	7	9
for names beginning:	Qa-Qt					
use numbers:	2-29					

4. After other initial consonants

for the 2nd letter:	a	e	i	o	r	u	y
use number:	3	4	5	6	7	8	9

5. When an additional number is preferred

for the 3rd letter:	a-d	e-h	i-l	m	n-q	r-t	u-w	x-z
use number:	2*	3	4	5	6	7	8	9

*(optional for 3rd letter a or b.)

Letters not included in these tables are assigned the next higher or lower number as required by previous assignments in the particular class.

As a general practice, the Library of Congress has added the publication date to the classification number and the author number. For example, a complete call number may look like this: Z693.W94 1991. Z693 is the LC classification number, representing the subject *cataloging*, .94 is the author number for Wynar, and 1991 is the publication date. In the MARC format, the LC call number is found in the field with tag number 050. For a locally assigned LC call number, 090 is the proper field.

When assigning classification numbers and Cutter numbers, the library technician must take time to read the instructions carefully and follow the directions precisely. It is both interesting and challenging. With practice and experience, the job of classifying material will fall into its logical sequence.

REVIEW QUESTIONS

1. What are the two major classification systems? How are they different?
2. Describe the four volumes of *Dewey Decimal Classification,* Twenty-first Edition.
3. In a Dewey library, what tools are used for assigning the complete call number? Explain and give an example.
4. In a library using the Library of Congress Classification system, how is a complete call number constructed? Explain and give an example.
5. What are the ten main classes of the Dewey system?
6. What are the letters used for classes in the Library of Congress Classification system?
7. How do the book numbers differ in Dewey and LC systems?
8. Outline the procedures for assigning Dewey numbers.

Chapter 7

Copy Cataloging

Now that we have learned about descriptive cataloging, subject cataloging, and classification systems, the process of cataloging is complete. Cataloging makes it possible for the users to determine, by checking under author, title, or subject, if the needed materials are contained in a particular collection. The call number on the card or displayed on the computer screen serves as a location guide, making it convenient for the user to go to a specific shelf to get the material. What we see on the card or screen is the result of the process of cataloging, called a bibliographic record, in library terms. When cataloging is done step-by-step in-house, as explained in the previous chapters, it is called original cataloging, and it is necessary for library technicians working in the cataloging department to have the knowledge required for original cataloging. However, the majority of new acquisitions have already been cataloged by some other cataloger at the Library of Congress or, perhaps, at another library, and there is no need to spend the time and effort to "reinvent the wheel." If such is the case, we adopt the cataloging already done by someone else for our own local use, by recording it without change, or by making minor changes or modifications to suit the local situation. The process of taking already established cataloging information and applying it for local use is called *copy cataloging,* that is, preparing a bibliographic record of our own by using or adapting the bibliographic record prepared by a cataloger from another library or organization. Copy cataloging is a widespread practice in the library world. It saves personnel, time, and money, resulting in speedier service while maintaining high quality. Copy cataloging should be performed whenever possible.

TERMINOLOGY

bibliographic utility: A consortium, or a network of automated libraries sharing one or more machine readable databases. Bibliographic utilities may be large or small, and they may be national, international, or regional in scope.

CIP: Abbreviation for Cataloging in Publication. In this practice, the cataloging information prepared by the Library of Congress before the book is published is incorporated into the book and printed on the copyright page (the back of the title page).

MARC: Refers to Machine Readable Cataloging. MARC tapes are computer tapes with bibliographic records, done in MARC format, compiled by the Library of Congress. MARC tapes are used for copy cataloging either through direct access to the tapes or by participating in a network that uses MARC tapes as a main source for its database.

network: A group of automated libraries that join together for the purpose of sharing information by using the same library application system from a centralized computer facility.

OCLC: Abbreviation for Online Computer Library Center. This is the world's largest and most comprehensive bibliographic utility, consisting of over 30,000 members throughout the world, and currently with a database of millions of records that is growing every day.

union list: A combined list of the holdings of many libraries. It may be on paper or, more likely, in electronic format, usually referred to as the online union list.

To perform copy cataloging, one must be able to find the record created by someone else. One way to do that is through the CIP information. Another is buying the records from commercial library service companies. In neither case is there a contribution to the database, which does occur, however, when copy cataloging is done by shared cataloging or cooperative cataloging, with the use of a bibliographic utility. The databases in bibliographic utilities usually include the MARC database from the Library of Congress plus the cataloging records created by participating libraries. For automated libraries that are part of a consortium, copy cataloging can be done by retrieving the record from the database, and, simultaneously, if the record is not in existence, creating a new record in the MARC format and adding it to the database.

Copy cataloging is done at different levels. One level is true copy cataloging, that is, duplicating a record so that it is an exact copy of the original work. The second level of copy cataloging is to use another library's record as a base and edit or modify the record to meet local needs. If the description of the record is an exact match, and if the originator is a reputable source such as the Library of Congress, usually the record is copied without any changes. On the other hand, if it is a so-called *near match,* or *close copy,* meaning that some information, perhaps the edition or the imprint, does not match what is found, then a decision has to be made. In some cases, a little editing or modifying will do. According to the OCLC standard followed by automated libraries, a new record must be created if the wording in the title is different; if it is a different edition; if the publication place or the publisher is different; if the publication date is different; if paging is different; and if the size has a difference of more than two centimeters. Careful judgment differentiates between a new record being created versus an old record being duplicated. Inserting records already there results in a dirty database. The library technician must be very familiar with a library's policies and must adhere to such policies when performing the task of copy cataloging.

SOURCES FOR COPY CATALOGING

The Library of Congress

The most extensive and commonly used sources for copy cataloging come from the Library of Congress in book form, on computer disk, or in CD-ROM format. The Cataloging Distribution Service (CDS) Division of the Library of Congress has the following publications, which are all sources for copy cataloging. For more information and costs, check the CDS Web page <lcweb.loc.gov/cds/union.html>.

In microfiche format, there are *National Union Catalog on Microfiche* and *The Music Catalog on Microfiche*. In print format, there are *National Union Catalog: Books, National Union Catalog: Audio*-visual Materials, and *National Union Catalog: Cartographic Materials*. All these publications cover bibliographic records of materials cataloged by the Library of Congress from 1983 to the most current year. References for earlier materials have the same title but with a different date as part of the title.

Figure 7.1 is a page reproduced from the *National Union Catalog*. For example, the book *Through Russian Eyes* by Gromyko can be found on this page. From the Library of Congress entry, you can copy not only information for descriptive cataloging but also the subject heading and classification number as well. Notice that both the LC number and the Dewey number are printed as part of the bibliographic record for the convenience of libraries using either system. The Library of Congress, which needs the LC number for itself only, adds the Dewey Decimal numbers to approximately two-thirds of the books as a service to the library community, helping to reduce the effort and cost for cataloging in Dewey libraries.

For Dewey libraries, the number after the decimal point is separated by an apostrophe mark ('), called a *prime mark*. This is done to meet the needs of libraries of various sizes. For example, Dewey number 973.922' 092'4 is listed for the book by Gromyko in Figure 7.1. This means that a small library with few books on this topic may use 973.922 as the classification number, while 973.922092 may be a more suitable number for a larger library having a larger collection on this topic. For an even larger library, or a library that has a large collection of historical materials, an even more specific number 973.9220924 may be the best choice. Note that the author number is not listed and therefore needs to be assigned by the local cataloger, either according to the rules of the Cutter table or any other local system the library may use. This Dewey number is used as a whole on the catalog card or on the computer database without the prime marks.

Automated libraries can subscribe to the Library of Congress cataloging records on MARC (machine readable cataloging) tapes. More often, libraries will share MARC tapes with other libraries through a consortium arrangement. When using the MARC database as a copy cataloging source, the copy cataloger matches the material on hand to

FIGURE 7.1. Sample Page from the *National Union Catalog*

Gromadzkie rady narodowe w świetle badań empirycznych przed reforma. Praca zbiorowa pod red. Sylwestra Zawadzkiego. Warszawa, Panstwowe Wydawn. Naukowe, 1973.
379 p. 24 cm. z154.00
Summary in English and Russian.
At head of title: Polska Akademia Nauk. Instytut Nauk Prawnych.
Includes bibliographical references.
1. Soviets—Poland. I. zowadski, Sylwester, ed. II. Polska Akademia Nauk. Instytut Nauk Prawnych.
JS6132.G74   74-207376

Groman, Margaret Winifred, 1944-
Bimultiplication rings and ring extensions. [Syracuse, N.Y.] 1972.
115 L.
Thesis—Syracuse University.
Vita.
Bibliography: leaf 115.
Microfilm of typescript. Ann Arbor, Mich., University Microfilms, 1972. 1 reel. (Doctoral dissertation series 72-20, 335)
NSyU   NYC74-40173

Groniowski, Krzysztof.
Polska emigracja zarobkowa w Brazylii, 1871-1914. Wroclaw, Zaklad Narodowy im. Ossolińskich, 1972.
296 p. 24 cm. z155.00
At head of title: Polska Akademia Nauk. Instytut Historii.
Summary in English.
Includes bibliographical references.
1. Poland—Emigration and immigration. 2. Brazil—Emigration and immigration. Poles in Brazil. I. Title
JV8195.G75   74-204172

Grontkowski, Christine Rosenbauer, 1939-
The priority of the scientific image: an investigation of Wilfrid Sellars' ontological commitments, by Christine R. Grontkowski. New York, 1969 [1972]
iii, 205 l.
Thesis—Fordham University.
Vita.
Bibliography: leaves [196]-205.
Photocopy of typescript. Ann Arbor, Mich., University Microfilms, 1972. 20 cm.
1. Sellars, Wilfrid. 2. Ontology. I. Title.
IEdS   NUC74-40163

Grosier, Jean Baptiste Gabriel Alexandre, 1743-1828.
The world of ancient China. Text by J. B. Grosier. Translated by Lana Castellano and Christiana Campbell-Thomson. Genève, Minerva, c1972.
144 p. illus. (part. col.)
Cover title: The ancient China.
1. China—Civilization. I. Title. II. Title: The ancient China.
CIU   NUC74-40177

Gross, G F
A revision of the species of Australian and New Guinea shield bugs formerly placed in the genera Poecilometis Dallas and Eumecopus Dallas (Heteroptera: Pentatomidae), with description of new species and selection of lectotypes. Melbourne, Australian Journal of Zoology, 1972.
192 p. (Australian Journal of Zoology. Supplementary series, no. 15)
Includes bibliography.
1. Heteroptera.
TxFTC   DNAL   NUC74-34431

FIGURE 7.1 (continued)

Groman, William A
Forest fertilization: A state-of-the-art review and description of environmental effects [by] William A. Groman. Corvallis, Or., Supt. of Docs., U.S. Govt. Print. Off., 1972.
57 p. 27 cm. (U.S. Environmental Protection Agency. Environmental protection technology series, 1972, no. 16)
I. Title. II. Series.
PSt CtY   NUC74-40188

Gromyko, Anatolii Andreevich
Through Russian eyes: President Kennedy's 1036 days, by Anatolii Andreevich Gromyko. [Authorized translation edited by Philip A. Garon].
Washington, International Library, 1973.
xviii, 239 p. 23 cm. $9.95
Translation of Tysiacha tridtsat' shest' dnef prezidenta Kennedi.
1. Kennedy, John Fitzgerald, Pres. U. S., 1917-1963. I. Title.
E841.G713 1973b 973.92'09'24 74-156772
                                          MARC

Groomes, Benjamin Herbert, 1934-
Study of the academic performance of students participating in an experimental curriculum as compared with students enrolled in the regular curriculum in the freshman and sophomore years of college at Florida Agricultural and Mechanical University, 1967 to 1969. [Gainesville, Fla.] 1971 [1972]
1v.
Thesis—University of Florida
Microfilm of typescript. Ann Arbor, Mich., University Microfilms, 1972. 1 reel. 35 mm.
1. Florida. Agricultural and Mechanical University, Tallahassee-Curricula. 2. Curriculum change. 3 Socially handicapped-Education.(Higher)-Tallahassee.
FMU   NUC74-40176

Groomes, Freddie Lang, 1934-
A study of human relations training designed to build a learning community. [Tallahassee] c 1972.
vi, 168 l.
Thesis (Ph.D.)—Florida State University.
Bibliography: leaves [157]-164.
1. Dissertations, Academic-F. S. U.—Education—Counselor education. 2 Interpersonal relations. 3. Group psychotherapy. 4. Group counseling. I. Title
FTaSU NUC74-40175

Gross, Gershon Wolfe, 1931-
An evaluation of labor relations in the dairy industry. [College, Park, Md.] 1972.
278 l. tables. 29 cm.
Typescript.
Thesis—University of Maryland.
Vita.
Includes bibliography.
1. Dairying. 2. Trade-unions. 3. Industrial relations. I. Title.
MdU   NUC74-34430

Gross, Joseph, 1934-
Company promoters, by Joseph H. Gross. [Tel-Aviv] Israel Institute of Business Research, Tel-Aviv University, 1972.
xxxix, 264 p. 25 cm. (Faculty of Law, Tel-Aviv University. Legal studies no. 2).
Label mounted on t. p.: Distributed in U.S.A. by F. B. Rothman, South Hackensack, N.J.
Includes bibliographical references.
1. Promoters. 2. Corporation law. I. Title. II. Series: Legal studies (Tel-Aviv) no. 2
346.06'62   74-155457
                                          MARC

Gross, Joseph F., joint author
see Gersten, Klaus. Transverse curvature effects in flows. . . Santa Monica, Calif., Rand, 1972.

the same title already in the database, cataloged by the Library of Congress. The local library code or symbol is then entered and the cataloging is done.

In 1996, the Library of Congress started a cooperative arrangement called Program for Cooperative Cataloging. Currently, the membership, made up of mostly libraries with large collections, totals 340. Membership contributes not only to the original MARC records for monographs (books) and serials; but also to name authorities, subject authorities, and classification proposals. As a result, the bibliographic records from the MARC tapes that are displayed on the screen may not have originated from the Library of Congress, but from a participating institution. Libraries universally benefit from this vast database of high-quality cataloging.

Cataloging in Publication (CIP)

In most books published in the United States, cataloging information can be found on the copyright page, which is the back of the title page. This is called Cataloging in Publication, a project started by the Library of Congress in 1971. Over one million CIP records have been processed, and 4,500 American publishers currently participate in this program. Under the agreement, before a book is published, the publisher sends the galley proofs, or galley surrogates, such as title page, copyright page, series page, table of contents, and sample chapters to indicate subject coverage, to the Library of Congress. The Library of Congress catalogs the material within ten days and sends the cataloging information back to the publisher so that the bibliographic record can be included and printed on the copyright page of the book when the book is published. This project benefits the libraries enormously because the information comes simultaneously with the book, and no further searching for cataloging information is necessary. Utilized properly, CIP is the best source for copy cataloging.

Because the cataloging is done before the book is published, information on the full description of the book, such as paging and other physical description, cannot be included. Also, when the book is published on a later date, the title, subtitle, and the date of publication may be changed. The copy cataloger must be very careful in checking all the elements when examining and transcribing the CIP information and make the necessary changes.

When a book with CIP information is published, the Library of Congress will upgrade the CIP information to the full cataloging level and replace the old, incomplete, temporary record with the new one. The LC staff compares the prepublication cataloging record with the bibliographic elements in the published book and makes necessary changes to ensure that the revised record accurately describes the publication and all the physical description elements added to the new record. All the CIP records and the subsequent verified records are entered as part of the MARC record database.

Recently, the Library of Congress instituted a program in which publishers can transmit the needed information electronically to the Library of Congress, and vice versa. Named ECIP, for electronic cataloging in publication, the project promotes efficiency and reduces paperwork, with libraries as the main beneficiaries.

To accelerate the upgrading of CIP records, the OCLC (Online Computer Library Center) has established an office at the Academic Book Center, staffed with OCLC catalogers. Recently, the OCLC has accepted the upgraded CIP records of the Yankee Peddler, a book dealer, as well. Upgrading records "on-site" at the book vendors has added another element of efficiency to the process and has resulted in widespread use.

Figure 7.2 is an example of CIP information reproduced from the copyright page of a book—the first edition of this textbook, *Cataloging and Classification for Library Technicians*. From the example, we can see that information on many elements needs to be added. Still, you will find the main and added entries, the subject headings, and the applicable classification number, all of which will speed your cataloging process.

Commercial Sources

If requested, commercial book vendors such as Bro-Dart and Baker & Taylor provide card sets at a minimum fee for all the materials ordered from them. The sets are complete with cataloging information and come with the books or other type of media. Also, such library

FIGURE 7.2. Example of Cataloging-in-Publication Data

The Haworth Press, Inc., 10 Alice Street, Binghamton, NY 13904-1580

Library of Congress Cataloging-in-Publication Data

Kao, Mary Liu

      Cataloging and classification for library technicians/Mary Liu Kao.
   p. cm.
      Includes bibliographical references and index.
      ISBN 1-56024-344-9 (acid free paper).
      1. Cataloging—United States. 2. Classification—Books.

I. Title.
Z693.U6K36   1995   94-44815
025.3'0973—DC20   CIP

service companies will process the materials on demand by including pockets and cards for each item, allowing immediate shelving after receipt by libraries. For small libraries with a limited staff and budget that use the manual system, this is the most economical way to complete the cataloging process. Commercial companies offer MARC databases on CD-ROM format that provide immediate, unrestricted online access, and without the costly connecting charges. This format is most attractive for small libraries and for libraries doing a large number of retrospective conversions.

OCLC (Online Computer Library Center)

A network, also called a consortium or a bibliographic utility, is an automated cooperative venture among libraries. Resources on the database may be entered by the central office staff or contributed by member libraries. The combined database is then transferred and shared by member libraries for reference, interlibrary loan, and cataloging purposes. Libraries belonging to a network can find on the database all the necessary cataloging information for a large percentage of their materials.

Many international, national, and regional networks exist. The information from all of them can be used for copy cataloging. Some networks also offer additional cataloging services to member libraries. OCLC is the most utilized.

With a membership of over 30,000 libraries in sixty-five countries, OCLC is the oldest and the largest library network. The OCLC database consists of MARC tapes that include all the cataloging done by the Library of Congress, plus cataloging contributed by the affiliate libraries. Libraries join OCLC through a regional broker network, such as NELINET (New England Library Information Network) in Boston for New England libraries.

OCLC and its shared form of cataloging have transformed cataloging practices across the United States. The cataloger uses OCLC's Online Union Catalog to locate cataloging information. With a database of millions of bibliographic records, it is estimated that over 80 percent of any library's cataloging needs can be filled, thus greatly reducing the time-consuming task of original cataloging. Since copy cataloging has become so common, traditional original cataloging is used only for the unique or unusual materials that cannot be found in the database. This change in emphasis has placed the library technician into the cataloging profession, which previously had been the librarian's domain.

In addition to its Online Union Catalog (OLUC) that is used throughout the world, OCLC offers some other cataloging services. One is the Bibliographic Record Notification service. This service automatically delivers online upgraded MARC records and additional features. OCLC's PromptCat service starts when the materials are being ordered. The vendor notifies OCLC of the titles ordered by the library, and OCLC finds the matching record in its Online Union Catalog, sets the library's holding symbol, and sends a copy of the MARC record to the library to load into the local system. In addition, it will send to the vendor electronic files of labels for physical processing. By the time the materials are received, the records already have been loaded and the materials are ready to be shelved. OCLC's TECHPRO service provides off-site contract cataloging and physical processing to meet a library's specifications. It is a customized cataloging service by contract, such as cataloging of foreign language materials. TECHPRO can be used to catalog special types of materials, such as Chinese-language books, or to clear up a backlog. The CatCD for Windows software offers Windows-based CD-ROM cataloging in an offline, stand-alone environment and gives the subscribing library access to a subset, CD-ROM only, of the OCLC Online Union Catalog. OCLC also offers a Retrospective Conversion service, which helps a library to convert all of its records to full MARC format. The OCLC CatExpress service offers Web-based copy cataloging for customers. OCLC discontinues services from time to time and adds new services, depending on the developing library trends and events. For the most current services that OCLC offers, check its Web site <oclc.org/oclc/menu/col.htm>. Up-to-date handbooks and operation manuals of all kinds are available, and the OCLC regional brokers provide workshops and on-site training for membership libraries so that library technicians can keep abreast of appropriate developments related to the system used in their libraries.

Figure 7.3 shows a MARC record as retrieved from the OCLC database. From this record, one can copy the call number, the subject headings, and other pertinent information from field 050, field 650, and so on. If copy cataloging online, for a total match, entering your library's code is the only requirement. The code DLC in field 040 means that the source of this record is the Library of Congress. The LC classification number designated by the Library of Congress is put in the 050 field.

Figure 7.4 is an OCLC record contributed by a member library. It contains information such as call number, subject headings, added entry, and the description of the material so that the library technician can perform copy cataloging quickly and easily. In this example, the MIA, MIA, and OCL in the 040 field show that this record is not inputted by the Library of Congress (DLC), but by a member library whose symbol is MIA (Miami University, Oxford, Ohio). The LC classification that is assigned locally is put in the 090 field. More details about fields in the MARC record are discussed in Chapter 8.

Other large national networks that are used the same way as described in the OCLC section are RLIN and UTLAS. Local and regional networks are also available as sources for copy cataloging, although their databases usually are not as large.

FIGURE 7.3. Example of MARC Record Contributed by the Library of Congress

---

Beginning of record displayed.   SID: 05572

OLUC dt get , to , kn , y   Record 10 of 161
    NO HOLDINGS IN MHG – 1 OTHER HOLDING
       OCLC: 29220370   Rec stat: a
       Entered: 19910809   Replaced: 19931030   Used: 19931030
    Type: a   Bib lv: m   Source: Lang: eng
    Repr: Enc lvl: Conf pub: 0   Ctry: xx
    Indx: 0   Mod rec: Govt pub: Cont:
    Desc: a   Int lvl: Festschr: 0   Illus: a
        F/B: 0   Dat tp: s   Dates: 1991,

1	010	91-90154
2	040	DLC c DLC
3	020	c $10.00
4	050 00	BX1756.M33985 b G48 1991
5	090	b
6	049	MHGA
7	100 1	Meade, Jim.
8	245 10	Getting to know you!/ c by Jim Meade.
9	260	[S.l.] : b J. Meade, c c1991.
10	300	106 p. : b ill. ; c 21 cm.
11	650 20	Church year sermons.
12	610 20	Catholic Church x Sermons.
13	650 0	Sermons, American.

Source: Copyright OCLC Online Computer Library Center, Incorporated 1997. Reprinted by permission.

The Internet

The Internet has provided another avenue for copy cataloging. Searching the online public access catalogs (OPACs) of other libraries via an Internet connection may yield useful information for copy catalogers. Library OPACs may be searched directly or through a search engine that is Z39.50 based, making it possible to search hundreds of library catalogs simultaneously. With special software, MARC records can be retrieved, displayed, and printed from these online catalogs. The editor in the program allows you to edit the chosen MARC records and transfer the records to your library system.

    Library technicians can retrieve current information from Web sites of the Library of Congress <www.lcweb.lox.gov> and OCLC <www.oclc.org>, both constantly updated. Other sources such as the

FIGURE 7.4. Example of MARC Record Contributed by a Member Library

---

SYNCOFO–⌐PRISM ⌐BLKJ ⌐ ⌐   ⌐ ⌐

Beginning of record displayed.   SID: 05572   OL

```
OLUC  dt  get , to , kn , y                      Record 83 of 161
      NO HOLDINGS IN MHG— 62 OTHER HOLDINGS
      OCLC: 18922040       Rec stat: c
      Entered:   19881219  Replaced:      19910204   Used: 19931201
      Type: 0    Bib lvlILL  m      Source: d    d      Lang:  eng
      Type mat:  b  Enc lvl:  l      Govt pub:        Ctry:  txu
      Int lvl:  b     Mod rec:        Tech:  n        Leng:  nn
      Desc:  a       Accomp:         dat tp: s       Dates: 1988,
 1  040      MIA c MIA d OCL
 2  007      v b f d c e b f a g h h o
 3  007      s b s d l e u f n g j h l i c n e
 4  090      QP111.6 b .G48
 5  090      b
 6  049      MHGA
 7  245  00  Getting to know your heart h kit : b lower elementary.
 8  260      Dallas, Tex. : b American Heart Association, c c1988.
 9  300      1 videocassette (VHS), 2 sound cassettes, 2 stethoscopes, activity
cards, alcohol prep packages, tubing, guides, worksheets, booklets ; c in box, 33 x 27 x
7 cm.
10  500      "The American Heart Association schoolsite program."
11  500      Intended audience: Grades 1-3.
12  520      Helps students learn the basics of heart-healthy living.
13  505  0   How your heart works—Smoking and your body—Food, fun and fitness.
14  650  0   Heart.
15  650  0   Cardiovascular system.
16  650  0   Health education (Elementary)
17  710  21  American Heart Association.
```

Source: Copyright OCLC Online Computer Library Center, Incorporated 1997. Reprinted by permission.

Internet Library for Librarians <www.itcompany.com/inforetriever/ and webCATs: Library Catalogues on the World Wide Web <sla.org/ chapter/ctor/toolbook/resource/index.html> offer information on cataloging and cataloging resources. Keep in mind that Web sites come and go. Professional journals and the online listserv for catalogers will provide up-to-date information in this area.

It is important to remember in doing copy cataloging that information can be adopted completely only if an exact match exists. Slight differences, such as different publisher, or different edition, different binding, mean no match. In such cases, information retrieved from the search must be altered and modified, and a new entry established.

As cataloging has evolved into a library technician's area of responsibility, education and training for that position has become more extensive. Knowledge in technical areas and familiarity with copy cataloging sources are essential. If copy cataloging information cannot be located, the library technician may have to perform original cataloging. In that case, it is necessary to apply the procedures involved with descriptive cataloging, subject cataloging, and classification, as discussed in Chapters 4, 5, and 6. Sound judgment must be exercised regarding what to adopt without change, what needs to be modified, and how the records should be modified, composed, or recorded.

REVIEW QUESTIONS

1. What is copy cataloging?
2. Should libraries perform copy cataloging whenever possible? Why?
3. For libraries belonging to computer networks, how is copy cataloging done?
4. For libraries using a manual card system, how is copy cataloging done?
5. List five sources for copy cataloging.
6. Why do more and more libraries hire library technicians instead of librarians as catalogers?

Chapter 8

Cataloging on Computers

TERMINOLOGY

delimiter: The character, or symbol, used to precede each subfield in the MARC format. Depending on the software, different symbols are used, but the most common are $, ≠, and _ .

field: On the MARC format, the bibliographic information of each record is separated into different parts, such as author, title, publisher, etc., forming fields in which each different part has a separate line with a number attached to the line. Field corresponds to the term *area,* as described in AACR2R.

indicator: A one-digit numeric code that follows the tag and precedes the field. For some fields, two numbers are used, one giving instruction to the computer for processing information, and the other providing information about the content of the field. The two are independent numbers.

leader: The leader is the first twenty-four characters of the MARC record. Information contained in the leader includes record status, type of record, bibliographic level, and others. These are all fixed fields, and the information is for computer use only.

MARC: MARC stands for machine readable cataloging. It consists of a system of inputting the cataloging information on the computer following a standard devised by the Library of Congress. This standardized format allows other libraries to share the data, the computer to interpret the data, and users to retrieve the data. MARC has been adopted by bibliographic utilities and individual libraries that catalog on computers.

MARC 21: A continuation of USMARC and CAN/MARC (MARC standards used by Canadian libraries) that is a harmonized format of the two, with differences between the two formats eliminated.

PREMARC: The 4.2 million older records (pre-1968) filmed from the Library of Congress shelflist and converted to the MARC format by Carrolton Press. Also called REMARC.

subfield: Each field in the MARC record contains several related pieces of data, and each piece of data is called a subfield. Each subfield is preceded by a delimiter, such as $ or ≠ , and by a subfield code, such as *a* or *b*.

tag: A string of three-digit numbers indicating the different fields, such as 100 for author, 245 for title.

USMARC: The MARC records created, used, and distributed by the Library of Congress. Libraries in the United States follow its specified standards. Originally called LC MARC in the 1960s, and USMARC until 1999, MARC 21 is the revised identification.

INTRODUCTION

Now in the twenty-first century, most libraries either have converted or are converting to computer cataloging. The card catalog is no longer a common sight. As a library technician whose job involves cataloging, chances are that you will be working at a computer terminal. Your work will consist of copy cataloging mostly, but occasionally you will have to do some original cataloging. When performing copy cataloging, you will pull up records from the database that match the materials you are cataloging and simply add your library's symbol to them. With original cataloging, however, you will pull up a blank MARC work form on the computer and input data into each field, according to the specified rules explained in Chapters 4, 5, and 6. Because the information to be recorded via the computer is the same as that which would be typed onto a catalog card, and because the method of extracting information from the item remains the same, the rules applied will follow AACR2R for descriptive cataloging, *Sears* or LCSH for subject heading, and LCC or DDC for classification number. As in the card environment, where pieces of information are arranged on the card, in the computer environment, the necessary information requires its proper place. The Library of Congress MARC format provides the structure that is followed.

THE MARC FORMAT

All automated libraries follow the MARC format. Figure 8.1 demonstrates how a MARC record is organized, by fields with tag numbers attached. This chart will provide you with some general knowledge of what a MARC record includes and how it compares to the information placed on the catalog card.

Within each hundred group, much information can be expressed by the "XX". For example, within the 1XX field for main entry, there may be a 100 field for personal author, a 110 field for corporate name, or 111 field for conference as main entry. In the 2XX field, there may be 240 for uniform title, 245 for title and statement of responsibility, 250 for edition statement, and 260 for imprint statement. These three-digit *tag* numbers tell the computer what kind of information is to follow.

FIGURE 8.1. Fields in a MARC Record

0XX	Control information, such as the identification number, classification number, etc.
1XX	Main entry, usually name of the author
2XX	Title and statement of responsibility
3XX	Physical description, etc.
4XX	Series statement
5XX	Notes
6XX	Subject headings
7XX	Added entries
8XX	Series added entries
9XX	Reserved for local use

In Chapter 4, you learned about areas and elements that need to be recorded on the catalog card. With the MARC format, the elements within the areas, the main and added entries, plus subject headings and other relevant information are all presented in separate lines in logical order. Each line starts with a three-digit number called a *tag*, followed by two single (separated) numbers, called *indicators,* and then the descriptive phrases, called *fields,* and finally *subfields,* which give a more detailed explanation of the fields. Each subfield is preceded by a subfield code, which consists of a delimiter and a lowercase letter. A more detailed explanation with examples is demonstrated in Figure 8.2.

These are the most commonly used fields, with the preceding tag numbers that a cataloger needs. For example, tag 245 tells the computer that title and statement of responsibility will follow, and the two indicators (_ _) contain important information about the item. The first indicator 0 would mean no title added entry is needed, while 1 would mean title added entry is needed. The second indicator tells the computer how many nonfiling characters there are, ranging from 0 to 9. For example, if a title starts with article *the,* the second indicator will be 4, which instructs the computer to skip four spaces (t, h, e, and space) when filing this title. A few more commonly used indicators are listed in Figure 8.3.

Figure 8.3, which is adapted from the MARC format published by the Library of Congress, demonstrates how indicators are assigned to lead to a more refined description. When cataloging online, you should consult the OCLC manual or the manual used by your bibliographic utility, which provides all MARC fields.

Recalling Figure 8.2, after the indicators come the fields, which include author, title, or subject areas. A sample of fields and subfields are listed in Figure 8.4 to demonstrate how they are organized.

Every field has numerous subfields. A guide to the MARC format, either published by the Library of Congress or by your bibliographic utility such as OCLC, should be kept handy when cataloging online. Publications such as *MARC Format Reference Card: Books* and *MARC Format Reference Card: Serials* are excellent resources. Very rarely will you need most of the tags mentioned here. According to the Library of Congress, only 10 percent of the tags are used frequently, the other 90 percent are used only occasionally.

FIGURE 8.2. Commonly Used MARC Tags and Fields

020	:	ISBN number
040	:	Cataloging source
050	:	LC call number
082	:	Dewey call number
090	:	Locally assigned LC call number
092	:	Locally assigned Dewey call number
100 _ _	:	Personal name as main entry
110 _ _	:	Corporate name as main entry
111 2 0	:	Conference name as main entry
240 1 0	:	Uniform title
245 _ _	:	Title and statement of responsibility
250 _ _	:	Edition
260 _ _	:	Imprint
300 _ _	:	Physical description
440 _ _	:	Series traced
490 _ 0	:	Series not traced
490 1 _	:	Series traced differently
500 _ _	:	General note
502 _ _	:	Dissertation note
504 _ _	:	Bibliography note
505 _ _	:	Contents note
520 _ _	:	Annotation note
533 _ _	:	Photoreproduction note
600 _ _	:	Personal name as subject
610 _ _	:	Corporate name as subject
630 _ _	:	Uniform title as subject
650 _ _	:	Topical subject
651 _ _	:	Subject with geographical area as first element
700 _ _	:	Personal name as added entry
710 _ _	:	Corporate name as added entry
711 2 0	:	Conference as added entry
730 _ _	:	Uniform title as added entry
740 _ _	:	Title added entry
830 _ _	:	Series added entry
856 _ _	:	Electronic location and access

FIGURE 8.3. Commonly Used MARC Indicators

Tag	Field	Indicator
100	Main entry, personal author	1: 0-forename only, 1-surname, 3-name of family 2: 0-not subject, 1-also subject
110	Main entry, corporate name	1: 0-surname inverted, 1-place name, 2-name in direct order 2: 0-not subject, 1-also subject
111	Main entry, conference name	Same as 110
240	Uniform title	1: 0-not on LC card, 1-on LC card 2: nonfiling characters (0-9)
245	Title statement	1: 0-no title added entry needed, 1-title added entry needed 2: nonfiling characters (0-9)
260	Imprint statement	1: 0-publisher present, 1-publisher not present 2: blank
300	Physical description	1: blank 2: blank
440	Series statement, title	1: blank 2: nonfiling characters (1-9)
505	Contents note	1: 0-complete, 1-incomplete, 2-partial 2: blank
600	Subject added entry, personal name	1: 0-forename only, 1-surname, 3-name of family 2: 0-LC subject heading, 2-children's subject heading, 8-Sears subject heading
650	Subject added entry, topical	1: 0-no level of subject term specified, 1-primary term, term, 2-secondary term 2: same as 600
651	Subject added entry, geographic	1: blank 2: same as 600
700	Added entry, personal name	1: same as 100 2: 0-alternative entry, 1-secondary entry, 2-analytical entry
710	Added entry, corporate name	1: same as 110 2: same as 700
711	Added entry, conference name	1: same as 111 2: same as 700
730	Added entry, uniform title	1: nonfiling characters (1-9) 2: same as 700
740	Added entry, title traced differently	1: nonfiling characters (1-9) 2: same as 700
800	Series added entry, name/title	1: same as 100 2: blank

FIGURE 8.4. Commonly Used MARC Subfields

Tag	Fields and Subfields
010	LC card number $a-LCCN, $z-cancelled LCCN
020	International Standard Book Number $a-ISBN, $b-binding information, $c-terms of availability
040	Cataloging source $a-original cataloging other than LC, $b-language of cataloging code, $d-modifying agency code
049	Local holdings $a-holding library code, $c-copy statement, $l-local processing data
050	LC call number $a-classification number, $b-item number
082	Dewey Decimal Classification number $a-DDC number, $2-source (edition number)
090	Local call number $a-local classification number, $b-local item number
100	Main entry, personal name $a-name, $b-numeration, $c-titles, $d-dates of birth, death
245	Title statement $a-short title/title proper, $b-reminder of title, $h-media qualifier
260	Imprint statement $a-place of publication, $b-name of publisher, $c-date of publication
300	Physical description $a-physical description, $b-illustration, $c-size/height
440	Series statement, title traced $a-title, $n-number of part, $v-volume number
520	Summary, abstract, annotation, scope, etc. note $a-summary, etc. note, $z-source

Figures 8.5, 8.6, and 8.7 show a few examples illustrating how a bibliographic record looks when cataloged online using the MARC format. The examples are taken from the Library of Congress publication *Understanding USMARC Bibliographic*. Check <http://lcweb.loc.gov/marc/> for more detailed information.

A look at these records line by line will enhance your understanding of the organization of MARC records. First observed are several lines called the *leaders* in the fixed fields, up to the 008 field. These fixed fields, or the controlled variable fields, contain record data such as record status, type, the date the record was entered, the language, bibliographic level, etc., all supplied by the MARC system. For example, in Figure 8.8 the top part before the 040 field is the *leader*. You may have to change or add some data when they are different from the default set by the system, such as country code or language code.

From the 010 field on are the variable fields that were explained in the beginning of this chapter. Notice that only the relevant fields are

FIGURE 8.5. Example of Book Cataloged in MARC Format

Leader	01041cam	2200265 a 4500	
Control #	001	89048230 /AC/r91	
Control # Identifier	003	DLC	
DTLT	005	19911106082810.9	
Fixed Data	008	891101s1990 maua j 00110 eng	
LCCN	010 bb	‡a 89048230 /AC/r91	
ISBN	020 bb	‡a 0316107514 :	
		‡c $12.95	
ISBN	020 bb	‡a 0316107506 (pbk.) :	
		‡c $5.95 ($6.95 Can.)	

For sale by the Cataloging Distribution Service, Library of Congress, Washington, D.C. 20541, to which inquiries on current availability and price should be addressed.

Cat. source	040 bb	‡a DLC	
		‡c DLC	
		‡d DLC	
LC Call No.	050 00	‡a GV943.25	
		‡b .B74 1990	
Dewey No.	082 00	‡a 796.334/2	
		‡2 20	
ME:Pers Name	100 1b	‡a Brenner, Richard J.,	
		‡d 1941-	
Title	245 10	‡a Make the team.	
		‡p Soccer :	
		‡b a heads up guide to super soccer! /	
		‡c Richard J. Brenner.	
Edition	250 bb	‡a 1st ed.	
Publication	260 bb	‡a Boston :	
		‡b Little, Brown,	
		‡c c1990.	
Phys Desc	300 bb	‡a 127 p. :	
		‡b ill. ;	
		‡c 19 cm.	
Note: General	500 bb	‡a "A Sports illustrated for kids book."	
Note: Summary	520 bb	‡a Instructions for improving soccer skills. Discusses dribbling, heading, play-making, defense, conditioning, mental attitude, how to handle problems with coaches, parents, and other players, and the history of soccer.	
Subj: Topical	650 b0	‡a Soccer	
		‡x Juvenile literature.	
	650 b1	‡a Soccer.	
AE:Dif titl	740 01	‡a Heads up guide to super soccer.	

FIGURE 8.6. Example of a Video Recording Cataloged in MARC Format

---

```
Leader          *****ngm  22*****1a 4500
001          89711816 /F
003    DLC
005    19891107152635.3
007    vfucbaho
008    890719s1988    cau010 c    v1eng c
010    ƀƀ      ǂa    8911816 /F
020    ƀƀ      ǂc For sale ($195.00) or rent ($50.00)
040    ƀƀ      ǂa AIMS Media
050    10      ǂa TH9148
082    10      ǂa 613.6  ǂ2  11
245    04      ǂa The Adventures of Safety Frog. ǂp Fire safety ǂh [video
                  recording] /
                ǂc Century 21 Video, Inc.
260    ƀƀ      ǂa Van Nuys, Calif. :  ǂb AIMS Media,  ǂc 1988.
300    ƀƀ      ǂa 1 videocassette (10 min.) :  ǂb sd., col. ;  ǂc 1/2 in.
500    ƀƀ      ǂa Cataloged from contributor's data.
538    ƀƀ      ǂa VHS.
521    ƀƀ      ǂa Elementary grades.
530    ƀƀ      ǂa Issued also as motion picture.
520    ƀƀ      ǂa Safety Frog teaches children to be fire safe, explaining that
                  smart kids never play with matches. She shows how smoke
                  detectors work and explains why they are necessary. She
                  also describes how to avoid household accidents that lead
                  to fires and how to stop, drop, and roll if clothing catches fire.
650    ƀ0      ǂa Fire prevention ǂx Juvenile films.
650    ƀ0      ǂa Fire detectors ǂx Juvenile films.
650    ƀ0      ǂa Dwellings ǂx Fires and fire prevention  ǂx Juvenile films.
650    ƀ0      ǂa Puppet films.
650    ƀ1      ǂa Fire prevention.
650    ƀ1      ǂa Safety.
710    21      ǂa Century 21 Video, Inc.
710    21      ǂa AIMS Media.
740    01      ǂa Fire safety ǂh [video recording]
```

---

used more than once. For instance, in Figure 8.5, because there are two separate editions for this book, field 020 for the ISBN number is used twice. Field 650 also appears twice because two topical subject headings are assigned to this book. For the video recording record in Figure 8.6, more fields are used as compared to fields used for books only. Subfields for the physical description are usually more for non-

FIGURE 8.7. Example of a Computer File Cataloged in MARC Format

---

Leader   *****cmm   22***** a 4500
001   93022553
003   DLC
005   19930731163011.2
008   930305s1993   mnun   d   b   eng
010   ƀƀ   ǂa   93022553
020   ƀƀ   ǂa 0792902637 : ǂc $59.00
037   ƀƀ   ǂa A-336 ǂb MEC
040   ƀƀ   ǂa DLC ǂc DLC ǂd DLC
050   00   ǂa QA95
082   00   ǂa 513 ǂ2 12
245   00   ǂa Amazing arithmetricks ǂh [computer file].
250   ƀƀ   ǂa Version 1.0.
256   ƀƀ   ǂa Computer programs.
260   ƀƀ   ǂa Minneapolis, Minn. : ǂb MECC, ǂc c1993.
300   ƀƀ   ǂa 2 computer disks ; ǂc 3 1/2-5 1/4 in. + ǂe 1 manual (71 p.)
538   ƀƀ   ǂa System requirements: Apple II series; 128K RAM; floppy
                    disk drive; color monitor recommended.
500   ƀƀ   ǂa Title from title screen.
500   ƀƀ   ǂa Ed. statement from disk label.
500   ƀƀ   ǂa Copy-protected.
500   ƀƀ   ǂa Same software on both disks.
521   2ƀ   ǂa 5-12.
520   ƀƀ   ǂa Using the motivating environment of a carnival setting, stu-
                    dents are challenged to solve a variety of puzzles. Intended
                    to improve logic and mathematical problem solving skills. In-
                    cludes a provision allowing students to choose a level of diffi-
                    culty matched to ability.
650   ƀ0   ǂa Mathematical recreations.
                  ǂx Juvenile software.
650   ƀ0   ǂa Problem solving ǂx Juvenile software.
650   ƀ1   ǂa Mathematical recreations ǂx Software.
650   ƀ1   ǂa Problem solving ǂx Software
710   20   ǂa Minnesota Educational Computing Corporation.
753   ƀƀ   ǂa Apple II.

---

book materials because a more detailed description is needed. For computer materials in Figure 8.7 special field 538 for system require-ments is necessary. For cataloging Internet resources, field 856 for electronic location and access is used. This field calls for an active link to a World Wide Web site of which the URL address is displayed in the subfield $u. As shown in Figure 8.8, the MARC format for a serial is

FIGURE 8.8. Example of a Serial Cataloged in MARC Format

001		9477925
005		19981013194313.0
008		830504c19839999miuqr1p bo 0 a0eng dcas a
010		83645580	zsn 83000625
012		3	b3	i8403	k1	11
022	0	0737-8831
042		lc	ansdp
049		BIIA
050	00	Z671	b.L699
069	1	SR0051211
070	0	Z671.L69
072	0	X200
210	0	Libr. hi tech
222	0	Library hi tech
245	00	Library hi tech
246	10	Library hi tech
260		[Ann Arbor, Mich. :	bPierian Press, + c1983-
265		Pierian Press, P.O. Box 1808, Ann Arbor, MI 48106
300		v. :	bill. ;	c28 cm.
310		Quarterly
362	0	[Vol. 1, no. 1] (summer 1983)-
500		Title from cover
510	1	Book review index	x0524-0581	b1984-
510	1	Library literature	x0024-2373
510	2	Electronics and communications abstracts journal (Riverdale)	x0361-3313
510	2	ISMEC bulletin	x0306-0039
510	2	Library & information science abstracts	x0024-2179
510	2	Microcomputer index	x8756-7040	b1985-
650	0	Library science	xTechnological innovations	xPeriodicals
650	0	Libraries	xAutomation	xPeriodicals
650	0	Information science	xPeriodicals
650	2	Library Automation	xperiodicals
650	2	Library Science	xperiodicals
650	2	Information Science	xperiodicals
890		Library hi tech
901			cSer
936		Summer 1983	av. 2, no. 2

similar to a book, except for field 310 for "current frequency"; field 362 for dates of publication and volume designation"; field 555 for "cumulative index/finding aids note"; field 780 for "continues," when this is a new serial title replacing an old one; field 785 for "continued by," which requires the new title of the same serial; etc.

The precise structure, along with some subjective decisions, requires a full knowledge of online cataloging, making it necessary to understand thoroughly the MARC format. These numbers and codes are not to be memorized, but knowledge of how to apply them is most important. When cataloging online, you will have a handbook to guide you. The Library of Congress published several cataloging aids that serve as references for the MARC format. They are for sale by the CDS (Cataloging Distribution Service) division of the Library of Congress. The complete USMARC package includes the following publications: *USMARC Concise Formats; MARC 21 Format for Bibliographic Data; USMARC Format for Authority Data; USMARC for Holdings Data; USMARC for Classification Data; USMARC Format for Community Information; USMARC Code List for Languages; USMARC Code List for Countries; USMARC Code List for Geographic Areas; USMARC Code List for Relators, Sources, Description Conventions; USMARC Code List for Organizations;* and *USMARC Specifications for Record Structure, Character Sets, and Exchange Media.* For further information on these publications, check the Web page of the Library of Congress CDS division <http://lcweb.loc.gov/cds/marcdoc.html>.

ORIGINAL AND COPY CATALOGING ON COMPUTERS

When cataloging online, it is necessary to use your library's computer database to pull up the bibliographic record that matches the item you are cataloging. This can be accomplished in several different ways. With many access points to use, correct data are readily available. In most systems, a good start is the ISBN number, the LCCN number, the author, the title, and some other combinations of two or more elements. Also, in most systems, you may qualify the search, such as limiting it to material type, such as video recording, or by date, so that you do not need to browse inefficiently through a lot of titles to find your item. In the OCLC system, you may also search using a variety of special methods. One example is the 4, 3, 1 rule, whereby a user can enter the first four letters of the author's last name, add a comma, then the first three letters of the author's first name, a comma, and the first letter of the author's middle name. Another way to search is by title, using the first three letters of the first word, a comma, and then the first two letters of the title's second word, a comma, and then the first two letters of the title's third word, a comma, and finally the first letter of the fourth word in the title. This is referred to as the 3, 2, 2, 1 rule. OCLC searches employ other combinations and other systems use different search techniques. Most important are understanding the basics of the MARC format, referring to the handbook or manual of the system your library uses, and learning the detailed rules and procedures needed to retrieve the records.

Let us suppose that you find a record that matches exactly the item you are seeking to catalog. At this point, add your library's holding symbol to it and the cataloging process is done. Your system manual will tell you how to enter the symbol, which in many systems automatically is displayed on the screen, then press the *send* key to complete the copy cataloging process. If the record differs only slightly from your item, such as having a different publisher, a different edition, or is in any other way different even though the author and title are the same, it is not considered a match. The information on the screen, however, can be used to build a new entry for the item you have. Sometimes you will need to add, delete, or modify data to suit your local needs.

If the record for your item is not found, original cataloging is required. In such a case, you will pull up the *work form* and fill in all the necessary information, carefully matching all your data to the fields and subfields of the MARC format. At this time not only is your knowledge of the MARC format essential, as discussed in this chapter, but you will also need the skills you learned from the previous chapters for assigning subject headings, classification numbers, and the rules for describing the item physically. Only after you have entered all the data can the record be called complete. Finally, your library's symbol is added to the record as in copy cataloging.

In summary, no matter what automated system your library uses, it is a version of the MARC format. As long as you have a thorough understanding of the MARC format, you can catalog on your system by following the procedures outlined in the user's manual for the system.

REVIEW QUESTIONS

1. Explain the elements of a MARC record.
2. How does the information on the MARC record differ from what is on a catalog card?
3. Why is it essential for library technicians to understand the MARC record?
4. List the procedures for copy cataloging online.

Chapter 9

The Cataloging Department

TERMINOLOGY

authority file: Files of authorized names, series titles, or subject headings used in a catalog. The file is checked when doing cataloging to ensure consistency in the form of names, series titles, and subject headings. The card catalog or the OPAC database is usually the authority file.

processing: The task of physically preparing the materials for the shelves. Procedures involved differ from library to library, depending on decisions made locally. Tasks include attaching the spine labels to materials, typing cards, attaching bar codes to materials, attaching date-due slips to materials, and stamping materials with the property stamp.

shelflist: A file, either in card or book format or in the computer database, arranged in order by call numbers, showing the library's holdings in shelf order. The shelflist is used usually by library staff only for staff functions such as collection development or inventory control.

ORGANIZATION

From library to library, the organization of the cataloging department varies. Always included, however, are the cataloging and classification plus the processing of all materials. In other words, the preparation of materials for shelves can be broadly divided into two groups: cataloging first, processing second. So far we have discussed cataloging completely, both original and copy cataloging. The next step is the processing of the cataloged materials.

THE AUTHORITY FILE

The purpose of cataloging and having a catalog is so that users can have easy access to a well-organized collection. The optimum use of a catalog depends on authentic entries and a clean database. This requires that the name of a person or organization that serves as an access point appears in the same form, so that all the materials authored by the same person or corporate body occur together on the screen or are filed together in the same section of the catalog. Establishing a list of all the authoritative access points is called *authority control*. Access points that need to have authority control include name, series titles, and subject.

The guidelines to use for assigning authority records are in a Library of Congress publication titled *USMARC Format for Authority Data*. In this publication is the format of standard forms for names of persons and organizations, series titles, uniform titles, and subject terms used in bibliographic records. The chosen forms as the authoritative names were selected by the Library of Congress and the Name Authority Cooperative (NACO), which is composed of library representatives who also contribute to the authoritative name lists. The Library of Congress books database serves as the name authority file, and the *Library of Congress Subject Headings* is the subject authority file that catalogers should use. For online cataloging in the MARC format, fields 1XX, 7XX, and 4XX need to be checked against the authority files to ensure proper usage.

Since authority control is a time-consuming, labor-intensive task, many libraries, especially the automated ones, contract this work out to authority service vendors. The authority service vendor checks and revises all the new headings in the bibliographic records and provides new authority records for the libraries as well. In this case, access points no longer need to be checked individually and routinely by the in-house catalogers, and as a result, cataloging can be done more quickly.

CATALOGING ROUTINES

After the materials are received and properly checked in, they are sent to the cataloging department, where the first step of cataloging begins. The library technician first searches for cataloging information from the identified print sources or matches the titles with the existing titles in the database. If information for copy cataloging is not available, then the library technician will perform original cataloging, the details of which were discussed in Chapters 4, 5, and 6. At this point, information is entered into the MARC work form with the library's code attached, and the cataloging is done. Libraries that continue to use cards no longer produce the card sets on-site. Instead, commercially produced cards are used to save staff time and to ensure the quality of the production.

Please be aware that every library has its own rules and practices for cataloging routines. Most procedures manuals list tasks in order as follows:

1. Materials are received.
2. Search database or other identified sources for cataloging information.
3. If information is found, do copy cataloging.
4. If cataloging information is not found, do original cataloging.
5. For libraries with OPAC, enter all information onto the screen.
6. For libraries with cards, prepare and produce cards or buy cards from commercial vendors.

The number of cards needed is determined by the number of added entries: a card for each subject heading, one main entry card, and one shelflist card. Most often libraries will buy not only the card sets from commercial vendors but also the complete processing service, such as spine labels, pockets, cards, etc. For libraries that produce their own cards, a computer program such as *The Librarian's Helper: The Professional Cataloging Program* should be used. This type of program is used on a stand-alone computer and is designed to produce catalog cards in conformity with AACR2R standards, and to print labels, pockets, and cards. Such programs as the OCLC's Cataloging Label Program allow the cataloger to print labels from the text files.

Cataloging routines are completed in many ways, and library technicians should follow the practices of their own libraries.

Automation has changed the cataloging scene dramatically. With members sharing cataloging information from the same database, each contributing cataloger has to adhere completely to the rules and the standards, leaving no room for individual interpretation. The result of extensive use of online cataloging systems has been that the majority of the cataloging done in the library now is copy cataloging. This is different from the manual system, which requires that more original cataloging be done. Staff composition in the cataloging department has shifted also, with the library technician now the department's backbone.

PROCESSING ROUTINES

After the cataloging steps are completed, materials are processed so that they can be shelved in their proper places. The processing routines differ from library to library. In general, the following steps may be included:

1. Mark shelflist card (if still used) with proper identification symbols, such as accession number, bar code number, or copy number.
2. File shelflist cards in order by call numbers.
3. File catalog cards (if still used) in the card catalog according to *ALA Filing Rules, 1980.*
4. Attach pockets, cards, and date-due slips to materials.
5. Stamp materials with the library property stamp on designated pages or places.
6. Place plastic jacket on book.
7. Prepare and attach spine labels.
8. Attach and scan in bar code labels.
9. Attach security strips.

Automation simplifies the processing routines by eliminating steps such as typing, filing, and attaching pockets and cards to the materials.

FILING

As noted earlier, in automated libraries no filing needs to be done. The computer is programmed so that all access points are indexed and, therefore, retrievable. Libraries with cards to file in the catalog should follow the rules in *ALA Filing Rules, 1980,* published by the American Library Association. Some of the most basic and most commonly used rules are summarized in Figure 9.1. Figure 9.2 shows examples from the *ALA Filing Rules, 1980.* When in doubt, this reference should be consulted.

Though there may not be shelflist cards to maintain, it is important to learn how the call numbers are filed in order, since this is the way materials are shelved. Dewey call numbers are filed by classification numbers first. For the same classification numbers, the author numbers are compared, first alphabetically, then by decimal number. The following is an example of a correct sequence:

010	010	010.01	010.1	010.1	010.1
.A19	.A2	.A131	.A121	.A13	.A13
				Ca	Cr

Library of Congress call numbers are filed by letters first, then by numbers, followed by author numbers and other work marks. The following is an example of a correct sequence:

H	HA	HA	HA	HC	HV	HV
35	35	35	35.1	34	4291	4291
.A39	.A4	.A4	.A123	.A9	.A234	.A234
1965	1989	1993	1992	1993	.C689	.C71
						1998

Keep in mind that all author numbers follow a decimal point, even though the decimal points mistakenly may not have been printed on spine labels or on cards. Therefore, for two items with the same class number 010, author number .A19 is filed before .A2.

FIGURE 9.1. Most Commonly Used Filing Rules

---

1. All character strings beginning with numerals are arranged before character strings beginning with letters. (Rule 1)
2. Punctuation and all nonalphabetic signs and symbols are ignored. (Rule 1.2)
3. The ampersand (&) is filed as spelled out. (Rule 1.3, optional)
4. Names and titles are interfiled, character by character. (Rule 2.1)
5. For records having identical access points, the order is references for main and added entries, main and added entries interfiled, references for subject entries, subject entries. (Rule 2.2)
6. Abbreviations are arranged exactly as written, not as spelled out. (Rule 3)
7. Initial articles that form an integral part of place name and personal name headings are regarded. Initial articles in the nominative case are ignored at the beginning of the access points. (Rule 4)
8. Initials, initialisms, and acronyms separated by spaces, dashes, hyphens, diagonal slashes, or periods are regarded as separate words. If only separated by other marks or symbols, or not separated in any way, they are regarded as single words. (Rule 4)
9. File numeric character strings according to numerical significance from lowest to highest. (Rule 8.1)
10. Punctuation used to increase the readability of a numeral is treated as if it does not exist. Punctuation used in other ways is treated as a space. (Rule 8.2)
11. Numerals after a decimal point are arranged digit by digit, one place at a time. Decimal numerals that are not combined with a whole numeral are arranged before the numeral 1. (Rule 8.3)
12. Characters in fractions are arranged in the following order: numerator, line (equal to space), denominator. (Rule 8.4)
13. Numerals in nonarabic notation are interfiled with their arabic equivalents (XIV = 14). (Rule 8.5)
14. Superscript and subscript numerals are filed as "on the line" numerals and preceded by a space. (Rule 8.6)
15. In a chronological file, dates are arranged according to chronology. (Rule 8.7.1)
16. A historic time period that is expressed only in words is treated as if it consists of a full range of dates for the period (16th century = 1500-1599). Geologic time periods are arranged alphabetically. (Rule 8.7.2)
17. Words that show the role of a person or corporate body in relation to a particular work are disregarded. (Rule 9)
18. In access points beginning with a surname, all terms of honor and address are disregarded. In access points other than those beginning with a surname, terms of honor and address are regarded. (Rule 10)

FIGURE 9.2. Examples Adopted from the *ALA Filing Rules,* 1980

EXAMPLES
Brown, John, 1610?-1679
Brown, John, 1610-1680
Brown, John, 1696?-1742
Brown, John, 1715-1766
Brown, John, 1800-1859
Brown, John, b. 1817

EXAMPLES
1:0 für Dich
1:00 a.m.
1,2-dithiolenes
1-3/4 yards of silk
1.3 acres
1, 3-cyclohexadienes
1^3 is 1
1/3 of an inch of French bread
1-bicyclobutylcopper (1) compounds
1 o'clock jump
1. Transfer RNA conformation . . .
2-1/2% PDQ interest tables
2′, 3′ isomeric specificity
2.5 percent
II-VI semiconducting compounds
2.8% interest
2+ and 3 !states in the even tin isotopes

EXAMPLES
London and Londoners
London, Andrea
London as it is today
LONDON BRIDGE
London bridge is falling down
London Conference on . . .
LONDON (CRUISER)
London, Declaration of, 1909
LONDON (DOG)
LONDON (ENGLAND)–ANTIQUITIES
London (England). Conference on . . .
London (England). County Council.
LONDON (ENGLAND)–DESCRIPTION
LONDON (ENGLAND : DIOCESE)
London (England). Guild Hall
London (England). International
  Conference on . . .
LONDON (ENGLAND)–POLITICS
  AND GOVERNMENT
London (England). Royal School of Mines
London (England). Symposium on . . .
London (England). University
London, Jack
London (Ky.)

EXAMPLES
EGYPT–HISTORY
– TO 332 B.C.   [0-332 B.C.]
– TO 640 A.D.   [0-640 A.D.]
– 332-30 B.C.
– GRAECO-ROMAN PERIOD, 332 B.C.-640 A.D.
– 30 B.C.-640 A.D.
– 640-1250
– 640-1882

UNITED STATES–HISTORY
    – COLONIAL PERIOD, CA. 1600-1775
    – KING WILLIAM'S WAR, 1689 -1697
    – QUEEN ANNE'S WAR, 1702-1713
    – FRENCH AND INDIAN WAR,
    1755-1763
    – REVOLUTION, 1775-1783
    – CONFEDERATION, 1783-1789
    – 1783-1815
    – 1783-1865
    – CONSTITUTIONAL PERIOD,
    1789-1809
    – 1801-1809
    – WAR OF 1812
    – WAR WITH ALGERIA, 1815
    – 1815-1861
    – CIVIL WAR 1861-1865
    – 1865-
    – 1865-1898
    – 1865-1921
    – WAR OF 1898
    – 1898-
    – 20TH CENTURY   [1900-1999]
    – 1901-1953

EXAMPLES
I-90 design team
I-95 harbor crossing corridor study
I., A.
I. A.A.
I.A.G. Literature on automation
I.A.M. Symposia on Microbiology
I am a mathematician
I and CS: the magazine of instruments . . .
I.B.   ["see" reference]
    I., B
    Brief discovrs dedie av Roy . . .
I.B.R.O.   ["see also" reference]
I built a bridge, and other poems
I.C.A. Congress

The job of filing, according to the *ALA Filing Rules 1980*, is completely eliminated in computerized libraries that have closed the card catalog and use only the online public access catalog (OPAC).

Routines in maintaining the card catalog include replacement of worn and soiled cards, adding *see* and *see also* reference cards as necessary, shifting drawers and changing labels accordingly, and adding guide cards for easy section identification. Maintaining a database involves keeping it up to date, with new bibliographic records added, new authority files loaded, status of materials clearly marked, etc. Other miscellaneous works done in the cataloging department include mending and repair, preservation of materials, keeping statistics, and following procedures for withdrawing titles to be discarded. It should be emphasized again that every library is different, and the library technician working in the cataloging department has to follow the policies and procedures of that individual library.

A catalog or a database is a living thing. New entries are entered every day, and at the same time, titles are deleted or cards are pulled out constantly. Accuracy is important, and one should exercise great care when filing cards or entering any information into the database, be it the correct spelling, the correct field, or the correct indicator. One card misfiled is one record lost forever. The same situation occurs in the computer environment. When the bibliographic information is not entered properly into the database, the record may be buried. Cataloging has to be done with knowledge and precision, and that is why the task is both interesting and challenging.

REVIEW QUESTIONS

1. List the procedures for copy cataloging in an automated library.
2. List the procedures for original cataloging in a manual library.
3. What is a shelflist file and the purpose of having a shelflist file?
4. How did automation change the cataloging department?
5. What is the changing role of the library technician in the cataloging department?
6. What reference tool is used for filing catalog cards?
7. What is the best way for a manual library to speed up the cataloging and processing tasks?

Chapter 10

Issues and Trends

The cataloging department of a library always has special projects to anticipate. Because of automation, practices and procedures have changed and copy cataloging has become the primary responsibility of the library technician. Unlike the card catalog which hides mistakes, the online catalog immediately exposes all mistakes, and, therefore, maintenance of the database is a challenging task. To ensure that the database is one of high quality, great care must be exercised when matching or re-creating records on the computer.

Another consequence of automation is the closing or freezing of the card catalog. Converting a library's collection to a machine readable format is an enormous job, especially for libraries with big collections. A library may decide to follow the new changes of the cataloging rules or to switch to a different classification system. The library technician may be involved in these projects and participate in special assignments. For any project, policies and procedures must be spelled out clearly, and in-house training should be provided for the library technician. The following are some of the most common issues that libraries deal with at the present time.

RECLASSIFICATION

Starting in the 1960s and early 1970s, many libraries, especially academic libraries, decided to switch from the Dewey Decimal Classification system to the Library of Congress Classification system.

Although it is relatively convenient for a computerized library with an online public access catalog to reclassify its collection, it is a time-consuming, labor-intensive project. Bibliographic records have to be pulled from the database and the classification numbers changed. The spine labels of the materials must be rewritten and the materials reshelved. Because of the cost, many libraries have reclassified only the most commonly used materials and the new acquisitions, resulting in libraries using both the Dewey and the Library of Congress Classification systems.

RECATALOGING

Maintaining the catalog is an ongoing, never-ending task. As the cataloging rules continually change, library catalogs, either online or card, need to be changed. Also, classification numbers and subject headings are updated constantly. Lost materials or any change of status must be reflected in the database and any errors discovered in the catalog need to be corrected. Besides the routine cataloging and processing, there is always some project to work on in the cataloging department. Each library has its own policy on how to deal with changes and new projects.

CLOSING THE CARD CATALOG

As automation becomes a common practice for many libraries, and as public access terminals are made available to the users, maintaining the card catalog becomes an inefficient use of both money and staff time, and as a consequence, the card catalog is not kept up to date and often is pronounced closed or frozen. After the total collection is online, many libraries discard the card catalog or keep the old catalog for occasional references. Closing the card catalog saves valuable staff time and money, eliminating filing and buying cards. Some libraries may choose to keep a card shelflist file for staff use, but, generally, the practice is discontinued because shelflists can be conveniently printed out for inventory purposes, etc.

RETROSPECTIVE CONVERSION

When libraries are automated, the new acquisitions are cataloged instantly online. For a library to circulate materials that were cataloged before automation took place, the information needs to be converted from catalog card to a machine readable format, a process referred to as *retrospective conversion*. Because retrospective conversion is a costly, time-consuming project, a library may contract the project to a commercial library service company, or the project may be done for a fee by the network or consortium of which the library is a part. An example of this service comes from OCLC, which performs conversion for individual libraries on a contractual basis, employing a step-by-step procedure for either the whole collection or for a special collection, such as foreign language materials. Libraries with a small budget may decide to complete the project in-house, with or without extra help. When attempted as an in-house project with no extra staff, a clear, step-by-step procedure manual for the retrospective conversion project needs to be in place.

OUTSOURCING

Throughout history, libraries have searched for cost-efficient ways to provide excellent services to users. One of the ideas and issues much talked about in the library world is *outsourcing*. Outsourcing means turning over the responsibility of a certain task to a commercial firm for a fee. Library cataloging is an appropriate target for outsourcing because the process is time-consuming and staff need special training to perform the job well. In the past, libraries have outsourced non-library-related services such as cleaning or accounting. For many years, to save costs, libraries have ordered cards from the Library of Congress and have contracted library vendors to supply cards and complete processing of materials. Outsourcing the whole cataloging operation or even the whole technical services department, and thus eliminating those departments completely, could be advantageous economically. However, high quality remains a concern.

COOPERATION

As more libraries share expenses, more efficient methods, and library materials, cooperation has become the mainstay. The networks of libraries are getting bigger. Many states now have consortia that include hundreds of libraries. OCLC has 30,000 members worldwide, and even the Library of Congress is seeking cooperative endeavors by establishing projects such as Program for Cooperative Cataloging (PCC), Name Authority Cooperative (NACO), and Cooperative Online Serials Program (CONSER). For more information, check the Library of Congress Web site <lcweb.loc.gov/catdir/pcc>. A milestone in cooperation in cataloging was reached in 1996 when the representatives of the Library of Congress and the British Library signed an agreement called Memorandum of Agreement on Convergence of Cataloguing Policy, paving the way for more seamless future interactions.

THE DUBLIN CORE

A present phenomenon is the proliferation of information available on the Internet, where often information cannot be retrieved easily. With the current indexing systems, mostly Boolean and word search systems, the search engines cannot possibly find all disciplines and cross-disciplinary information. Full cataloging for Internet materials is not feasible because of the time, effort, and expertise involved. Also, because the Internet information is so ephemeral, a full cataloging record is not warranted, as in a regular library collection. The Dublin Core represents the compromise for the situation. The goal is to define a core set of metadata elements (the counterpart of catalog data for printed materials) that will allow authors and information providers to describe their work and facilitate interoperability among resource discovery tools. The required elements include subject, title, author, publisher, other agent, date, object type, form, identifier, relation, source, language, and coverage. A wide variety of Internet information can be described using such a cataloging standard. As more information appears on the Internet, the library technician may be required to do cataloging according to the standard set by the Dublin Core. For more information on the Dublin Core, check <purf.org/metadata/dublin_core>.

TRENDS

What does the future hold for cataloging? As the individual user is able to retrieve more information by subject and by key word, it appears that the cataloging processes are becoming obsolete and that the cataloger may no longer be needed. On the contrary, the practice of cataloging will become more important because the organizational aspect of the ever-increasing body of knowledge and information is the essential basis of cataloging. A wealth of information stored in the computer may be rendered useless unless it is organized well enough to be retrieved quickly and easily. The procedures and practices may change, but the challenge for the cataloger remains more, not less, important in our information-oriented society.

Some phenomena, which already are developing or are predicted to happen, will appear on the scene in the cataloging world, including the following:

1. Libraries will be automated and will catalog online. Even the smallest libraries will be able to eliminate some parts of cataloging and processing chores, if only with a stand-alone computer system.
2. Libraries in the United States will have online public access catalogs and discard their card catalogs.
3. Libraries will join some kind of cooperative system, such as a regional, national, or international network, to facilitate cataloging and other related functions.
4. Print and nonprint materials will be cataloged with equal care, applying the same rules, and will be intershelved.
5. The role differences between copy catalogers and original catalogers will become blurred. Catalogers will start out doing copy cataloging and, if copy records cannot be found, will proceed with original cataloging.
6. With an automated system, key word search will become the most popular way of searching for information. More access points and less description of the materials will become the rule. The Dublin Core agreement is an example. AACR2R rules will be modified.
7. Small libraries will still use the CD-ROM database to catalog. Larger libraries or libraries with ample budgets will catalog online.
8. The Internet will be the union catalog of the world. More libraries will use the Internet as a source for copy cataloging.

9. Library technicians will be hired instead of librarians to do cataloging, making the library technicians' job more demanding and interesting, while keeping the cataloging cost down.

10. Integrated online systems will get more involved with centralized processing for member libraries.

11. Libraries will outsource the technical services operations, including cataloging, particularly cataloging of special collections.

12. Cataloging departments will be merged into the automation or bibliographic control departments because of the changing nature of the job.

13. Because library technicians are performing higher levels of cataloging, more education and training opportunities for library technicians will be provided. Continuing education for library technicians will be an important issue.

14. Libraries will move away from perfect cataloging, adopting full records from bibliographic utilities without editing or modification.

Automation in libraries began in the cataloging department and has had its biggest impact there. Library cataloging and catalogs have come a long way. From handwritten cards to typewriters, to duplicating facilities, to personal computers, to computer networks—library technicians have played, and will continue to play, a vital part in developing the new and innovative aspects of the library. No matter what the future holds, the essence of the cataloger's job, that of creating the link between the information and the user, has not and never will change.

REVIEW QUESTIONS

1. Define recataloging.
2. Define reclassification.
3. Why are many card catalogs closed?
4. Explain retrospective conversion.
5. Explain how automation has changed the cataloging department.
6. What is outsourcing? Why do many libraries outsource their cataloging activities?
7. List five future trends of the library world that are related to cataloging.

Suggested Readings

ALA Filing Rules. Chicago: American Library Association, 1980.

Anglo-American Cataloging Rules, Second Edition, 1998 Revision. Chicago: American Library Association, 1998.

Byrne, Deborah J. *MARC Manual: Understanding and Using MARC Records,* Second Edition. Englewood, CO: Libraries Unlimited, 1998.

Chan, Lois Mai. *Library of Congress Subject Headings: Principles and Applications,* Third Edition. Englewood, CO: Libraries Unlimited, 1995.

Chapman, Liz. *How to Catalog: A Practical Handbook Using AACR2 and Library of Congress,* Second Edition. London: Clive Bingley, 1990.

Crawford, Walt. *MARC for Library Use: Understanding Integrated USMARC,* Second Edition. Boston: G. K. Hall, 1989.

Crawford, Walt and Gorman, Michael. *Future Libraries: Dreams, Madness, and Reality.* Chicago: American Library Association, 1995.

Ferguson, Bobby. *Blitz Cataloging Workbooks.* Englewood, CO: Libraries Unlimited, 1998.

Fritz, Deborah A. *Cataloging with AACR2R and USMARC for Books, Computer Files, Serials, Sound Recordings, Videorecordings.* Chicago: American Library Association, 1998.

Gorman, Michael. *The Concise AACR2,* 1998 Revision. Chicago: American Library Association, 1998.

Hunter, Eric J. and Bakewell, K.G.B. *Cataloguing,* Third Edition. London: Library Association Publishing, 1991.

Intner, Sheila. *Special Libraries: A Cataloging Guide.* Englewood, CO: Libraries Unlimited, 1998.

Kascus, Marie A. and Hale, Dawn, eds. *Outsourcing Cataloging, Authority Work, and Physical Processing: A Checklist of Considerations.* Chicago: American Library Association, 1995.

Liheng, Carol and Chan, Winnie S. *Serials Cataloging Handbook: An Illustrative Guide to the Use of AACR2R and LC Rule Interpretations,* Second Edition. Chicago: American Library Association, 1998.

Maxwell, Robert L. and Maxwell, Margaret F. *Maxwell's Handbook for AACR2R.* Chicago: American Library Association, 1997.

Millsap, Larry and Ferl, Terry Ellen. *Descriptive Cataloging for the AACR2R and the Integrated MARC Format: A How-To-Do-It Workbook,* Revised Edition. New York: Neal-Schuman, 1997.

Olson, Nancy, ed. *Cataloging Internet Resources: A Manual and Practical Guide,* Second Edition. Dunblin, OH: OCLC Online Computer Library Center, Inc., 1997.

Rogers, Terry. *The Library Paraprofessional: Notes from the Underground.* Jefferson, NC: McFarland, 1997.

Saye, Jerry D. *Manheimer's Cataloging and Classification,* Fourth Edition. New York: Marcel Dekker, 1999.

Schultz, Lois Massengale. *A Beginner's Guide to Copy Cataloging on OCLC/PRISM.* Englewood, CO: Libraries Unlimited, 1995.

Taylor, Arlene G. *The Organization of Information.* Libraries Unlimited, 1999.

Warwick, Robert T. and Carlborg, Kenneth. *Using OCLC Under Prism.* New York: Neal-Schman, 1997.

Zuiderveld, Sharon, ed. *Cataloging Correctly for Kids,* Third Edition. Chicago: American Library Association, 1998.

Index

Page numbers followed by the letter "f" indicate figures.

Due to a composition error, index
page #145 is located on the back of
this page. All entries are included
and we apologize for this error.